Microprocessors and their Operating Systems

A Comprehensive Guide to 8-, 16- and 32-Bit Hardware, Assembly Language and Computer Architecture

APPLIED ELECTRICITY AND ELECTRONICS SERIES

Editor: P. Hammond, *University of Southampton, UK*

Pergamon Titles of Related Interest

DHEERE
Universal Computer Interfaces

HAMMOND
Electromagnetism for Engineers, 3rd Edition

HINDMARSH
Electrical Machines & Drives, Worked Examples, 2nd Edition
Electrical Machines & their Applications, 4th Edition

HOLLAND
Illustrated Dictionary of Microelectronics & Microcomputers
Integrated Circuits & Microprocessors
Microcomputers & their Interfacing

KUFFEL & ZAENGEL
High Voltage Engineering

YORKE
Electric Circuit Theory, 2nd Edition

Pergamon Related Journals (*free specimen copy gladly sent on request*)

Computers & Electrical Engineering

Electric Technology USSR

Microelectronics & Reliability

Robotics & Computer-integrated Manufacturing

Solid State Electronics

Microprocessors and their Operating Systems

A Comprehensive Guide to 8-, 16- and 32-Bit Hardware, Assembly Language and Computer Architecture

R. C. HOLLAND
West Glamorgan Institute of Higher Education, Swansea, UK

PERGAMON PRESS
OXFORD · NEW YORK · BEIJING · FRANKFURT
SÃO PAULO · SYDNEY · TOKYO · TORONTO

U.K.	Pergamon Press plc, Headington Hill Hall, Oxford OX3 0BW, England
U.S.A.	Pergamon Press, Inc., Maxwell House, Fairview Park, Elmsford, New York 10523, U.S.A.
PEOPLE'S REPUBLIC OF CHINA	Pergamon Press, Room 4037, Qianmen Hotel, Beijing, People's Republic of China
FEDERAL REPUBLIC OF GERMANY	Pergamon Press GmbH, Hammerweg 6, D-6242 Kronberg, Federal Republic of Germany
BRAZIL	Pergamon Editora Ltda, Rua Eça de Queiros, 346, CEP 04011, Paraiso, São Paulo, Brazil
AUSTRALIA	Pergamon Press Australia Pty Ltd., P.O. Box 544, Potts Point, N.S.W. 2011, Australia
JAPAN	Pergamon Press, 5th Floor, Matsuoka Central Building, 1-7-1 Nishishinjuku, Shinjuku-ku, Tokyo 160, Japan
CANADA	Pergamon Press Canada Ltd., Suite No. 271, 253 College Street, Toronto, Ontario, Canada M5T 1R5

First English edition 1989

Library of Congress Cataloging-in-Publication Data

Holland, R. C.
Microprocessors and their operating systems:
a comprehensive guide to 8-, 16- and 32-bit hardware,
assembly language and computer architecture
R. C. Holland.—1st ed.
p. cm.—(Applied electricity and electronics series)
Includes index.
1. Microprocessors. 2. Operating systems
(Computers). I. Title. II. Series.
QA76.5.H633 1989 004.16—dc 19 88-38877

British Library Cataloguing in Publication Data

Holland, R. C.
Microprocessors and their operating systems:
a comprehensive guide to 8-, 16- and 32-bit hardware,
assembly language and computer architecture
(Applied electricity and electronics)
I. Title II. Series
004.16

ISBN 0-08-037189-2 Hardcover
ISBN 0-08-037188-4 Flexicover

Printed in Great Britain by BPCC Wheatons Ltd, Exeter

CONTENTS

PREFACE

This book is an attempt to bring together all the essential features of each of the major microprocessor families that have evolved through the 1970s and 1980s. The main characteristics of each of the 8-bit, 16-bit and 32-bit microprocessors are presented, together with descriptions of applications and software considerations.

The book excludes some of the less-frequently applied microprocessors, but the concentration on the most common devices provides descriptions of the processors that are used in at least 95% of current microcomputer systems.

Most books on similar topics tend to concentrate on a single device. The comprehensive coverage of the subject presented here should provide a useful reference for the reader who requires a good general background to the whole subject. The later chapters in the book concentrate on programming languages (BASIC, Pascal and C), and in particular on operating systems (the main program in a multi-program microcomputer system). The descriptive notes on the facilities available within the most common operating systems should enable a user of a disc-based microcomputer to perform a complete range of operator functions.

An introductory chapter on microcomputer principles is provided to assist a reader who is new to the subject, and a glossary of terms at the end of the book should prove helpful.

The author wishes to thank several of his colleagues for helpful discussions, and his long-suffering students (past and present) who have often provided an experimental platform for the presentation style of much of the material.

CHAPTER 1

MICROCOMPUTER PRINCIPLES

1.1 Microcomputer Architecture

Computers are generally classified as follows:

(a) mainframe computers, which support a large number of user terminals and peripherals and which are used for large data-handling systems, e.g. payroll, customer accounts;
(b) minicomputers, which support a smaller number of user terminals (typically up to 40) and which are used in small-company business applications as well as factory control systems;
(c) microcomputers, which normally support a single user (although latest systems have edged into the minicomputer market) and which are applied in a wide number of domestic, commercial and industrial control applications.

The distinguishing feature of a microcomputer is that its CPU (central processor unit) is a single IC (integrated circuit). This central component is termed "microprocessor". It is supported by memory and input/output ICs to constitute a complete microcomputer.

The basic architecture of all types of computer is shown in Fig. 1.1. The CPU, or microprocessor in the case of a microcomputer, generates the three buses, which are a set of circuit connections. The computer program that the CPU executes is held in memory, which in microcomputers is invariably arranged in 8-bit (a bit is 0 or 1) locations. The CPU extracts each instruction in turn from memory, examines it and executes it. The transfer of each instruction is as follows:

(1) CPU sets the memory address of the next instruction on the address bus;
(2) memory location presents the contents of the selected memory address on the data bus;
(3) CPU reads in the instruction and implements it.

The control bus carries signals that are required to activate the above transfers.

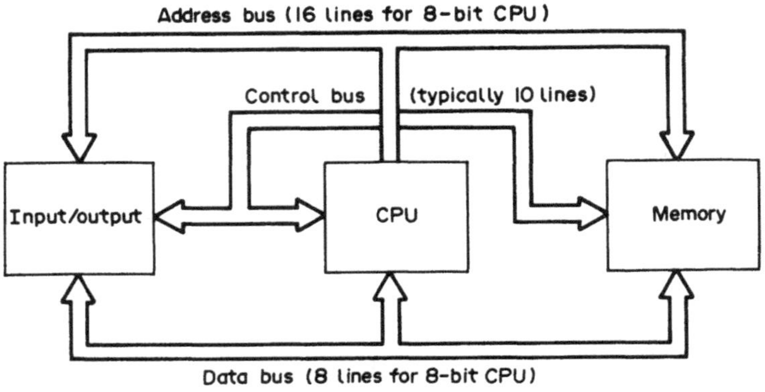

FIG. 1.1. Computer architecture.

Memory also contains data values in addition to program instructions. Data values can be numbers (represented in binary), or characters for display or printing purposes (represented by a unique code—the "ASCII" code).

The input/output module contains circuits that transfer programs and data to and from peripheral devices, e.g. backing store (such as floppy disc), printer and VDU (user terminal).

In the simplest forms of microcomputer, e.g. pocket calculator or washing machine controller, the three modules (CPU, memory and input/output) are contained within a single IC.

1.2 CPU (Central Processor Unit)

The typical internal organisation of a CPU is shown in Fig. 1.2. All actions are triggered by a clock pulse. The functions of the internal modules are as follows:

(a) The *program counter* contents are placed on the address bus so that the next program instruction can be transferred from memory to the CPU.
(b) The *instruction register* receives the instruction.
(c) The *control unit* decodes the instruction and activates transfers around the CPU in order to execute the instruction.
(d) The *work registers* hold data items that can be processed within the instruction (a specialised work register that can receive the results of ALU operations is termed an "accumulator").
(e) The *ALU* is the processing part of the CPU, i.e. it can alter data values using arithmetic (add, subtract, etc.) and logical (OR, AND, etc.) functions.

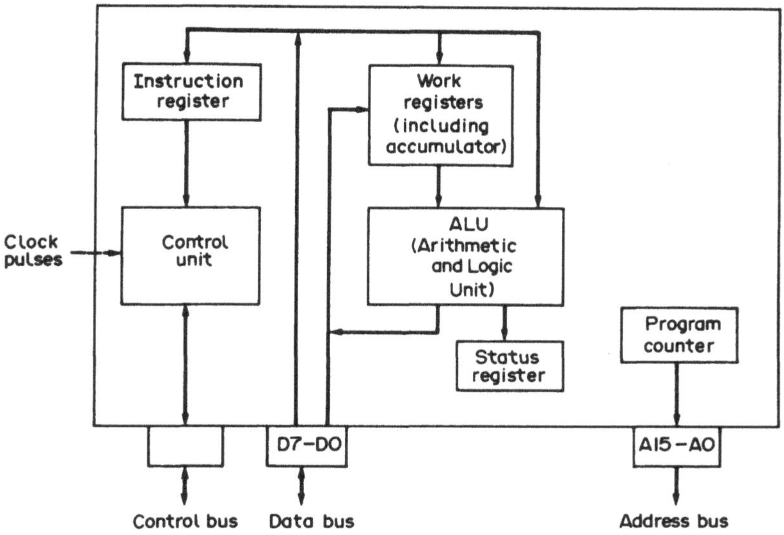

FIG. 1.2. Internal organisation of CPU.

(f) The *status register* contains "flags" that mark the result of ALU operations, e.g. zero result.

The word length of a CPU is expressed in terms of the number of bits that are processed in a single ALU operation, e.g. an 8-bit CPU can perform an arithmetic add operation on two 8-bit numbers. Microprocessor IC packages can be grouped as follows:

(a) 8-bit invariably 40-pin DIL (dual-in-line, i.e. pins on two sides only) package
(b) 16-bit normally 40- to 64-pin DIL package
(c) 32-bit between 60 and 120 pins in a "chip carrier" (pins on all four sides of package), or PGA (pin grid array).

The operation of each instruction is conveniently divided into two operations—"fetch" and "execute". Whilst the fetch part of the fetch–execute cycle is the same for every instruction, the execute part may or may not involve further memory transfers. Some instructions require only ALU operations in the execute part of the cycle, rather than slow memory transfer operations. Therefore the instruction time is variable and depends on the nature of the instruction. The following examples illustrate this:

Instruction Example 1—Add register B to register A

Step 1—Fetch instruction
Step 2—Execute (add B to A, the accumulator)

Instruction Example 2—Load register A from memory

Step 1—Fetch instruction
Step 2—Execute 1 (read first half of memory address into CPU)
Step 3—Execute 2 (read second half of memory address into CPU)
Step 4—Execute 3 (read data byte from memory address into A)

The second instruction takes approximately twice as long to operate as the first instruction. Notice that, in an 8-bit microprocessor, the first instruction is one byte (8 bits) long, i.e. it occupies one memory location. The second instruction is three bytes long, because the second and third bytes contain the 16-bit memory address of the data value that is being accessed. The second and third bytes in this case are termed the "operand" of the instruction, whilst the first byte that specifies the nature of the instruction is termed the "opcode".

The instruction set that can be obeyed by a CPU may number from 50 to several hundred different instructions. Instructions can be conveniently grouped into four categories, as follows:

(1) Data move, e.g. between a CPU register and another CPU register, a memory location or an input/output channel
(2) Data modify, i.e. when the ALU alters a data value
(3) Jump, i.e. transfer program control to another memory location
(4) Miscellaneous, e.g. to stop program execution.

Within (1) and (2) there are various ways of accessing a data value, and these are termed the "addressing modes" that can be specified within an instruction. The most common addressing modes are:

(a) Direct Register—data value is held in a CPU register
(b) Direct Memory—data value is held in a memory location
(c) Indirect Register—data value is held in a memory location specified within a CPU register (register-pair for an 8-bit CPU)
(d) Immediate—data value is held within the instruction (in the second byte for an 8-bit CPU)
(e) Indexed—similar to (c) except that a displacement within the instruction is added to the contents of the indexing register to compute the memory address.

1.3 Memory

Memory ICs are either ROM (read-only memory), which can only be read, or RAM (random access memory), which can be read from and written to. The pin connections of a typical ROM device are shown in Fig. 1.3. The 10 address lines give:

$2^{10} = 1024 \quad (1024 = 1K)$

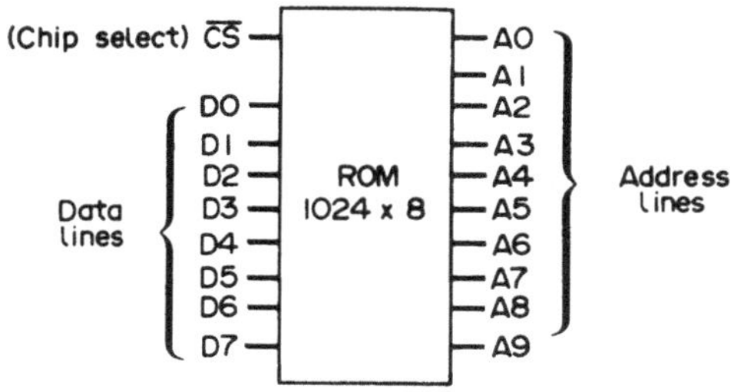

FIG. 1.3. Typical ROM.

memory locations, and the 8 data lines indicate that 8 bits are stored in each memory location. The "memory organisation" of the device is therefore declared to be:

1024 × 8

A byte is only read out of the device when the $\overline{CS}$ (Chip Select) signal is set to 0 (the bar above the CS legend denotes inverse logic, i.e. the signal is set when it is at logic 0).

The pin-compatible family of ROM devices is:

ROM—mask-prgrammed by the chip manufacturer
PROM—programmable ROM that is programmed by the user
EPROM—erasable PROM that can be erased using UV light and re-programmed.

The pin connections of RAM devices are similar, except that an additional CPU control bus signal (R/$\overline{W}$—Read/Write) must be connected to select the direction of byte transfer. There are two subdivisions of RAM:

Static RAM—holds its bit pattern until dc power is removed (i.e. it is "volatile").
Dynamic RAM—holds its bit pattern for only 2 ms unless a refresh circuit is applied

A complete microcomputer memory circuit is shown in Fig. 1.4. The ROM and RAM devices are connected to the data bus and also to as many of the least significant address bus lines as required. The Chip Select signals are generated by a "2 to 4 decoder", which ensures that only one memory device in the system can be selected at any time. The truth table for the decoder is shown in Fig. 1.5(a). Only one of the four outputs can be set to 0 at any time, and the $\overline{G}$ (Enable) input signal on the decoder must be set to 0 otherwise no

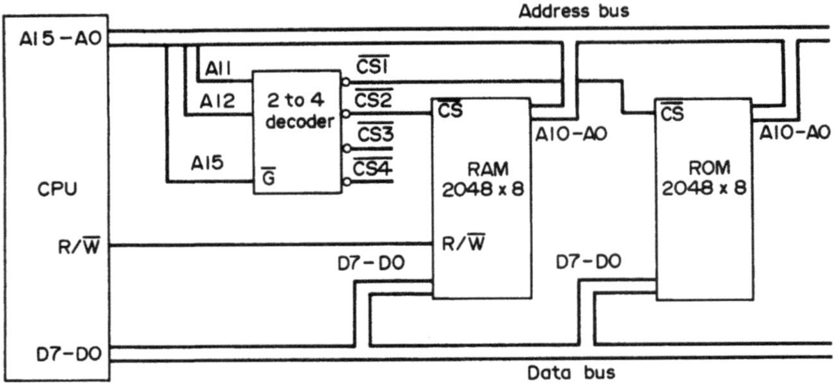

FIG. 1.4. Simple microcomputer memory circuit.

otherwise no output can be set. This particular memory circuit produces device addresses as follows:

	A15	A14	A13	A12	A11	A10	A9	A8	A7	A6	A5	A4	A3	A2	A1	A0	
ROM start	0	X	X	0	0	0	0	0	0	0	0	0	0	0	0	0	Hex 0000
ROM end	0	X	X	0	0	1	1	1	1	1	1	1	1	1	1	1	Hex 07FF
RAM start	0	X	X	0	1	0	0	0	0	0	0	0	0	0	0	0	Hex 0800
RAM end	0	X	X	0	1	1	1	1	1	1	1	1	1	1	1	1	Hex 0FFF
	Decoder enable			2 to 4 decoder													

X = not used (assume set to 0)

Grouping the 16 bits into 4 hexadecimal characters gives the addresses shown on the right of this table. These addresses can be shown on a "memory map" for the computer as indicated in Fig. 1.5(*b*), and this map serves as a guide to the computer programmer to indicate where he can place his programs and data lists.

A12	A11	$\overline{CS4}$	$\overline{CS3}$	$\overline{CS2}$	$\overline{CS1}$
0	0	1	1	1	0
0	1	1	1	0	1
1	0	1	0	1	1
1	1	0	1	1	1

(a) Truth table for 2 to 4 decoder in Fig. 1.4

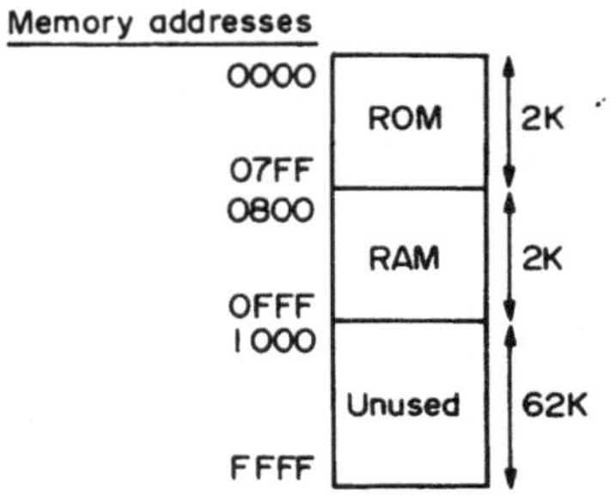

(b) Memory map

FIG. 1.5. Address calculations for Fig. 1.4.

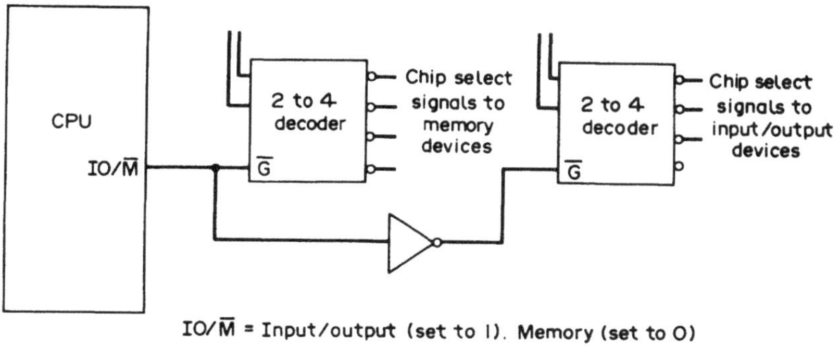

FIG. 1.6. Separate decoding circuits for memory and input/output.

If more than four memory devices exist in a memory circuit, a 3 to 8 decoder can be used in place of a 2 to 4 decoder, and the unused higher-order address bus lines must be connected at its inputs.

Several microprocessors do not possess input/output instructions, and input/output chips must be connected into circuit as if they are memory devices and addressed using memory transfer instructions. This arrangement is termed "memory mapped input/output". If a CPU does possess input/output instructions, memory and input/output devices must use separate decoders as shown in Fig. 1.6. In this arrangement memory and input/output chips can possess the same addresses, i.e. the memory map can overlap the input/output map. However no bus conflict can occur because the CPU sets the IO/$\overline{M}$ signal to 0 when performing memory transfers (to select the first decoder) and to 1 when performing input/output transfers (to select the second decoder).

1.4 Input/Output

Input/output chips are used to connect the microcomputer to a wide variety of peripheral equipment (backing store, printer, VDU), data links to other computers, instrumentation and electrical equipment in control applications, displays of various types, and other devices. Whilst memory chips (ROM and RAM) are supplied by a range of manufacturers who do not normally manufacture the CPU to which they connect, input/output devices are invariably tailored to the CPU and are offered as support devices by the CPU manufacturer.

The principal input/output chips are described in the following sections:

(a) PIO (Parallel Input/Output)

The function of a PIO is to pass data through 8-bit "ports" to and from remote devices. Figure 1.7 illustrates the pin functions of a typical PIO. The

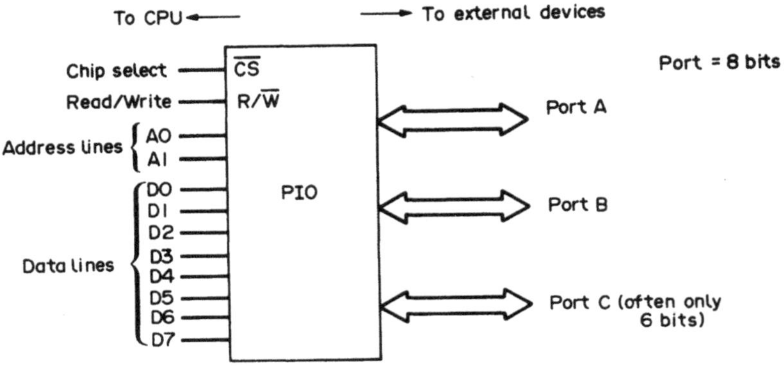

FIG. 1.7. PIO pin functions.

signal connections to the CPU buses are similar to that of a RAM chip. A data byte is transferred via the data bus, and typically two address bus lines are connected to provide four addresses on the PIO as follows:

A1	A0
0	0—Port A
0	1—Port B
1	0—Port C
1	1—Control register

The control register is used to select the directions of the ports, i.e. the programmer must place a control byte in the control register before bytes are transferred through the ports.

It should be noted that microprocessors, memory chips and programmable input/output chips are manufactured using the MOS (or CMOS) technology—this technology is often called VLSI (very large scale integration). However, a non-programmable alternative to a PIO is available in the form of an 8-bit register from the TTL family, which comprises ICs for gating and general digital handling circuits. A typical device that can be utilised as an input port is a SN74244, which possesses:

8 input lines —connect to external device/peripheral, e.g. 8 pushbuttons
8 output lines—connect to data bus
2 enable lines—one used as Chip Select and connected to address decoder

Typical applications of parallel output ports are:

(1) 8-bit connection to 7-segment display (+ decimal point)
(2) 1 to 8 bits to drive indicator LEDs/lamps or to drive electrical motor/heater/solenoid

(3) D/A (digital/analogue) converter, to produce an analogue signal for connection to a servo, graph plotter, control valve
(4) 8-bit connection to a parallel printer—the standard "Centronics" interface to a parallel printer comprises 8 data bits, 1 handshaking bit to printer, 1 handshaking bit from printer

Typical applications of parallel input ports are:

(1) A/D converter, to read an instrument signal (analogue) into a computer
(2) 1 to 8 bits from pushbuttons or switches ("contact-closure" signals).

(b) UART (Serial Input/Output)

The UART (universal asynchronous receiver/transmitter) is applied in place of a PIO when it is required to pass 8 data bits along a single serial connection in place of an 8-wire parallel connection. Although a reduction in cable cost is achieved, the disadvantage of this interfacing technique is the reduction in transfer speed. Serial transmission is applied for the connection of VDUs to computers, for inter-computer data links and for some printers.

Figure 1.8 shows the typical pin functions for a UART. The circuit connections to the CPU are similar to that of a PIO, except that a clock pulse signal is required to trigger transfer of the 8 serial data bits whenever a byte is transferred. The signal connections to the remote peripheral device, e.g. a VDU, are via a single Tx (out) and Rx (in) line. Four handshaking signals are also provided to ensure that the remote peripheral has finished processing the previous data byte (CTS) and is available for transfer (DSR). The signal interface to the serial-drive peripheral normally conforms to the RS232-C standard, and the pin connections required for the 25-pin interconnection plug/socket arrangement are shown on the right-hand side of the diagram.

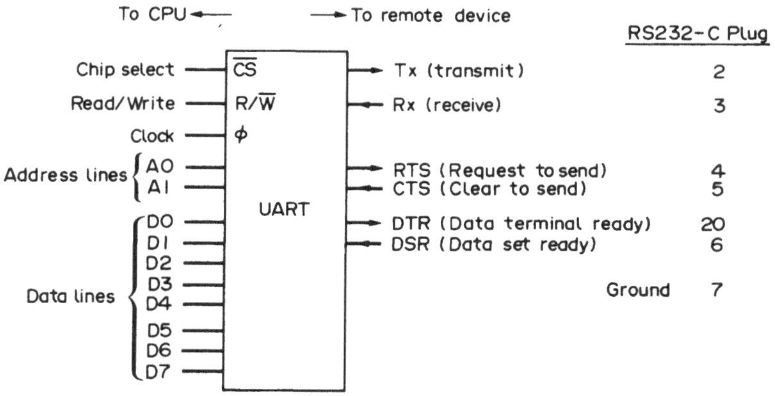

FIG. 1.8. UART pin functions.

There are typically four addresses on the UART, as follows:

A1	A0
0	0—Tx (serial out)
0	1—Rx (serial in)
1	0—Control register
1	1—Status register (e.g. character cleared, character received)

The 8-bit data character is framed by a start bit (logic 0) and a Stop bit (logic 1). The control register must be initialised by software, before data bytes can be transferred, to select the following options:

(a) serial transmission speed, measured in baud (bits/second)—typically 600, 1200, 2400, 4800, 9600 baud
(b) number of stop bits (1, $1\frac{1}{2}$ or 2)
(c) number of data bits (5, 6 or 7—normally 7 with a parity bit added by UART)
(d) parity (added after the data bits, and odd or even parity selected)

Notice that the UART operates at TTL voltage levels (logic 1 = +5 V, logic 0 = 0 V), but the RS232-C specification requires different voltage signals (logic 1 = −12 V, logic 0 = +12 V). Therefore a dc level shifter circuit is required between the UART and the interconnection plug/socket.

(c) CTC (Counter Timer Circuit)

A CTC (counter timer circuit) is a programmable counter that can be loaded, and read back to, the CPU. A CTC chip holds from one to four such counters typically. Sometimes a single counter is included within a PIO chip.

Figure 1.9 shows the three most common applications of a CTC channel (counter). In Fig. 1.9(*a*) the counter automatically reloads itself when the count reaches zero; in this way a regular stream of pulses is produced. In Fig. 1.9(*b*) an interrupt signal is generated which causes the CPU to execute a special interrupt program. This program can update counts in memory that represent the time-of-day, and other programs can either use this time-of-day clock or can be called at specific times. In Fig. 1.9(*c*) software can read the count achieved, and can also reset the counter to zero. Generally this is a more efficient way of counting external events, e.g. over a period of a day, compared with software responding to every event (perhaps by the generation of an interrupt).

Some CTCs can be programmed ("initialized") to either count up or count down. Additionally the incoming clock signal can be divided by programmable binary quantities before the counter is incremented/decremented.

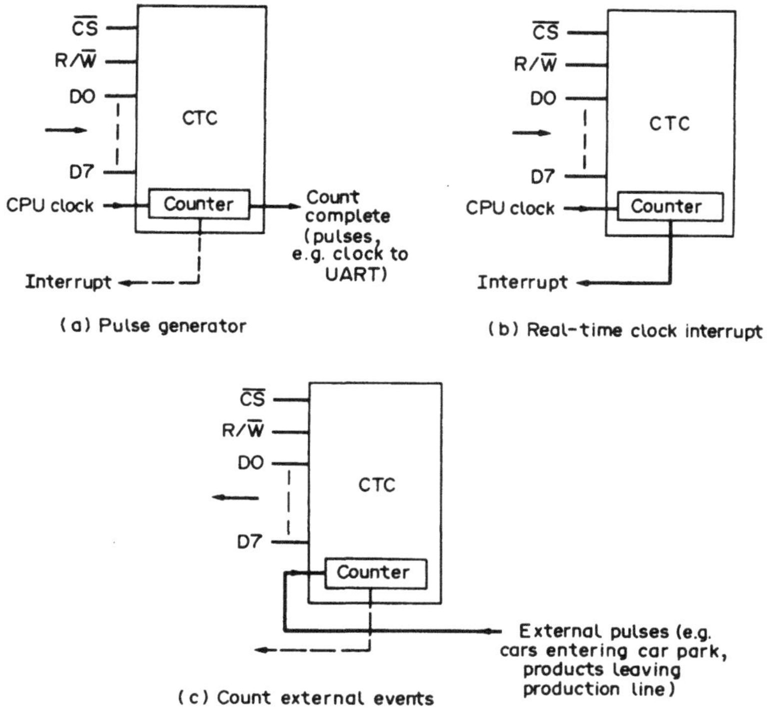

FIG. 1.9. CTC applications.

(d) FDC (Floppy Disc Controller)

The fourth programmable input/output device that is offered by most microprocessor manufacturers is the FDC (floppy disc controller), which can control one or two floppy disc drives. Figure 1.10 illustrates the typical circuit connections. The two address lines give four addressable registers on the chip, as follows:

A1	A0
0	0—Data register (for transfer of byte to and from disc)
0	1—Track register (to select required track)
1	0—Sector register (to select required sector)
1	1—Status/command register

A typical transfer procedure for a disc write operation is as follows. Software loads the track register (perhaps 1 of 77 concentric tracks) and sector register (typically 26 sectors per track, where 1 sector = 128 bytes) with the required position on the disc. The FDC moves the head assembly to the desired track

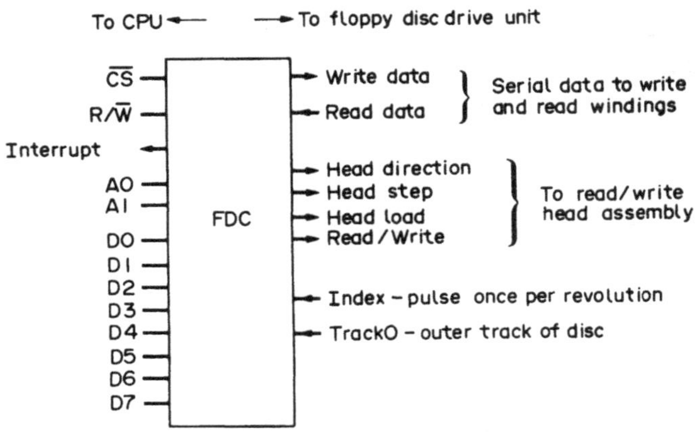

FIG. 1.10. FDC pin functions.

by pulsing the drive stepper motor, and waits until the disc has rotated until the required sector is beneath the head. The head is then "loaded" (placed against the disc surface) and the interrupt signal is generated. Software then loads the data register with the data byte to be written on the disc surface. This byte is fed to the read/write head ("write" selected) in serial form. Data byte transfers then repeat until the required block of data (typically several sectors) is transferred.

8″ floppy discs use the IBM3740 Format, which includes the writing of a sector self-address on the disc surface at the start of every sector. Alternative disc sizes are $5\frac{1}{4}''$ and $3\frac{1}{2}''$. Discs can be single-sided (SS) or double-sided (DS), and single-density (SD) or double-density (DD). Storage capacity is typically 500K bytes, and access time is variable and based on the following time components:

Disc revolution time	=	200 ms
Track-to-track time	=	8 ms
Head load time	=	35 ms
Total	=	35 ms to 1 s

A non-removable hard disc ("Winchester") offers 10M bytes, or more, and access times of 25 ms to 60 ms.

Note DMA (direct memory access) is a technique whereby input/output directly accesses memory, and transfers take place between the two without involvement by the CPU. An input/output circuit can request the CPU to release its buses (data, address and one or two control lines) by generating a "DMA Request" signal. When the CPU is not using the buses, it replies with a "DMA Acknowledge". The input/output circuit can then address memory directly and transfer bytes. This technique is often more efficient than the software alternative of transferring each byte under program control, and it is used frequently with backing store and video signal generation (for CRT displays) control purposes.

1.5 Stack, Subroutines and Interrupts

The "stack" is an area of memory (RAM) that is used to store return addresses when an exit is made from a main program to obey:

(a) a subroutine (activated by a CALL instruction in the main program)
(b) an interrupt service routine (ISR) when one of the interrupt lines (part of the CPU's control bus) is set.

Figure 1.11 shows the principle of operation of a subroutine and an interrupt service routine. A subroutine is a section of program that is normally placed at the end of a main program, and can be called several times from the main program. This arrangement makes more efficient use of memory than the

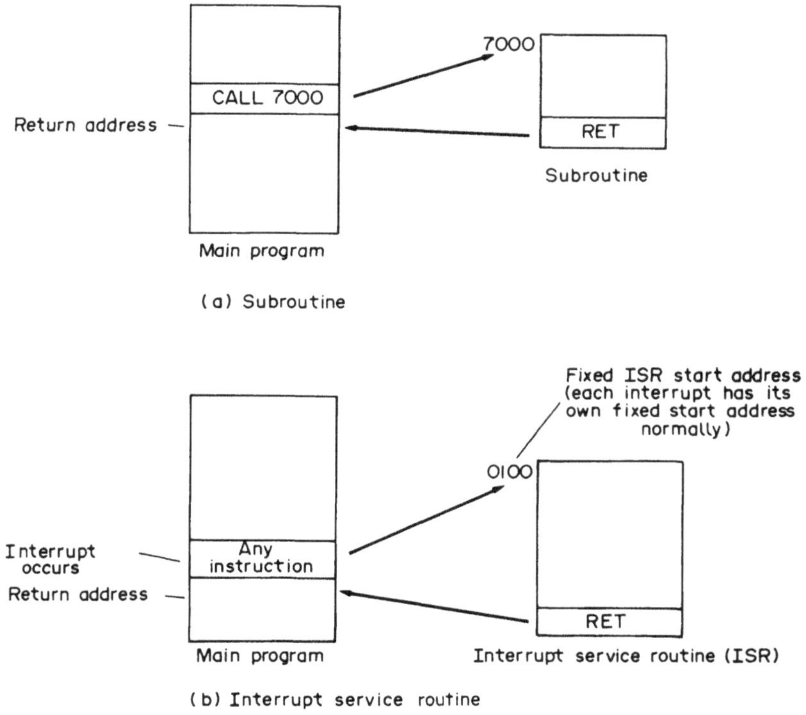

FIG. 1.11. Main program suspension by subroutine/interrupt.

alternative arrangement of placing that section of program at several positions in the main program. The CPU must store the address of the instruction that follows the CALL instruction (this is called the "return address"), so that this address can be reinstated in the program counter (see section 1.2) when the RET (return) instruction is obeyed at the end of the subroutine. The operation of an ISR is similar, except that an ISR is entered when a hardware signal (interrupt) is set. An ISR is normally a self-contained program that runs to completion, and also terminates with a RET instruction.

The mechanism of return address storage and retrieval is shown in Fig. 1.12, which assumes that an 8-bit microprocessor (e.g. the Zilog Z80 described in chapter 3) is used. The CPU's stack pointer is a 16-bit register that contains the memory address of the last used location on the stack. The first address in the main program contains an instruction that sets this register to a memory location in a free area of RAM. When the CALL instruction is executed the return address is placed on the stack in two separate memory locations, and the stack pointer is adjusted (decremented) by 2.

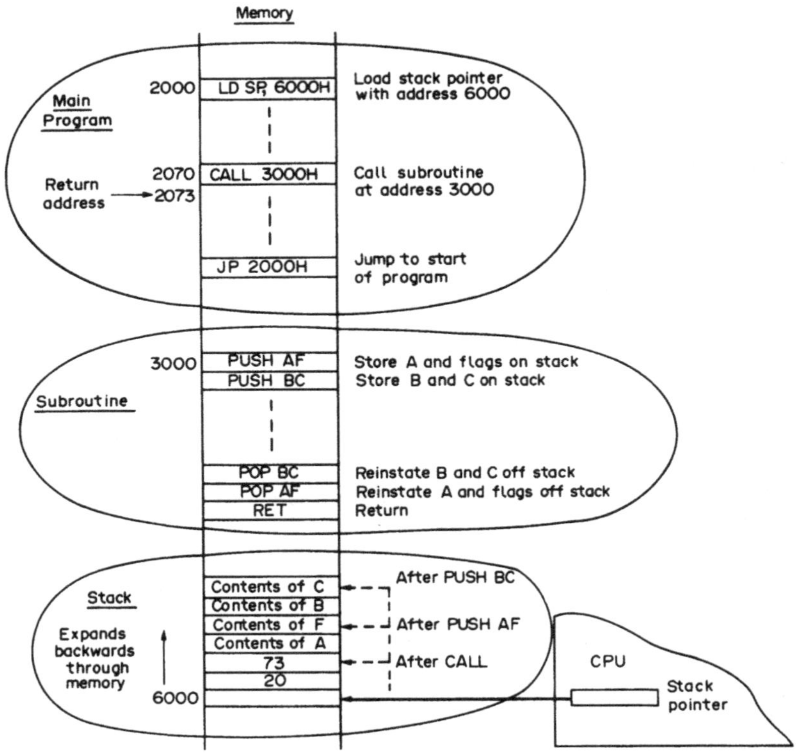

FIG. 1.12. Stack mechanism.

In this particular example the second application of the stack is illustrated.

The subroutine is likely to overwrite the CPU's work registers (A, B, C, etc., and perhaps the status register); an ISR is almost certain to do this. Therefore it is sensible to store the contents of these registers into memory locations at the start of the subroutine/ISR, and reinstate them at the end of the subroutine/ISR. The most convenient way of doing this is to use the PUSH (store registers on stack) and POP (reinstate registers off stack) instructions. The stack pointer is automatically adjusted after each PUSH and POP instruction.

A subroutine can call another subroutine in what is termed a "nested subroutine" system. In this case the second return address is placed on top of the first return address in the stack.

Different microprocessors use a variety of ways of specifying the start address of ISRs when interrupts occur. The most common methods are (PC = program counter):

(a) PC ← Fixed memory address
(b) PC ← "Vector address" (memory address contains start address of ISR)
(c) PC ← Vectored address, with address partly generated by interrupting device (e.g. Zilog Z80) and partly held within CPU

Microprocessors typically possess 3 or 4 interrupts. One interrupt (normally named "RESET") is arranged to become set when the computer system is switched on, or alternatively when a Reset pushbutton is pressed. In this way the system will always commence program operation at the same memory location—the most common start address is hexadecimal 0000, i.e. at the beginning of memory. In a disc-based computer this start-up program normally transfers the main program from disc into memory (RAM) and enters this program. This is called "bootstrapping" the system. The remaining interrupts may connect from "intelligent" input/output chips, e.g. CTC or FDC (see section 1.4). These interrupts normally possess a priority system, i.e. a lower priority interrupt cannot interrupt an ISR for a higher priority interrupt. Usually the CPU provides a "masking" system, such that a software instruction can be activated to block any further interrupts, even on a higher interrupt level.

Several microprocessors possess "software interrupts", or TRAP instructions, that have the same effect as hardware interrupts.

Bibliography

1. *Integrated Circuits and Microprocessors*. R. C. Holland. Pergamon, 1986.
2. *Microelectronic and Microprocessor-based Systems*. Noel M. Morris. Macmillan, 1985.

CHAPTER 2

INTEL 8080/8085 FAMILY (8-BIT)

2.1 Introduction

Intel were the first manufacturer to mass-produce an 8-bit microprocessor—the 8008—in the early 1970s. This 18-pin device was soon replaced with the 8080 and 8085 microprocessors, which proved to be the main stimulus to the widespread application of the microcomputer. Although neither of these microprocessors was used in the range of home computers that were produced throughout the 1970s, these devices dominated industrial control applications of the microcomputer during this period.

The operation of the 8080 and 8085 8-bit microprocessors is identical, except that an additional IC (the 8228) is required with the former to provide what is effectively a two-chip CPU. The facilities performed by this component are provided on-chip with the 8085, and comprise:

(a) on-chip clock generation
(b) system control (memory and input/output read/write operations) and interrupt prioritising.

2.2 8085 Pin Functions

The pin functions of the more popular 8085 are shown in Fig. 2.1. The most distinctive feature to notice is that the 8 data bus lines and the least significant 8 of the 16 address bus lines are multiplexed, i.e. they share the same pins. This is unique amongst 8-bit microprocessors, but is common amongst 16-bit microprocessors where requirements for interconnecting pin usage are more demanding due to the larger number of address and data bus signals. The pins in the 40-pin package that are saved by using this practice are available for use by the control bus and provide a powerful range of additional control functions. Examples are:

(a) In addition to a RESET signal, there are five interrupts (TRAP, RST 5.5, RST 6.5, RST 7.5 and INTR).
(b) Two serial input/output signals are available—SID (serial input data) and SOD (serial output data). Using particular instructions a single

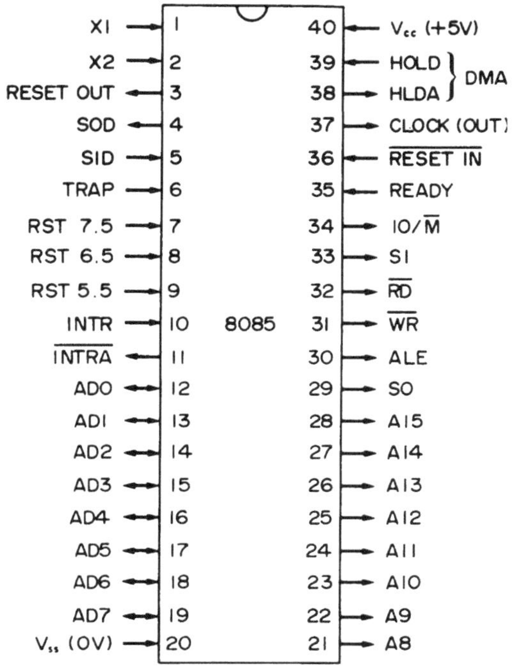

FIG. 2.1. 8085 pin functions.

bit can be transferred into and out of the accumulator using these signal pins.

Notice that only a crystal (or RC circuit) needs to be connected at the two clock input pins—the chip possesses an on-chip clock generator. The HOLD and HLDA signals are the DMA handshaking signals.

Intel support the 8085 with combined memory and input/output chips that have multiplexed data and address pins to provide connection compatibility with the 8085. However, if conventional memory chips are required in the circuit, then the two buses must be demultiplexed. This is performed by the addition of a further IC (e.g. 8212), and the CPU pin ALE (address latch enable) carries a control signal that identifies whether address or data is set on these lines (set high for address, low for data).

2.3 Processor Architecture

The internal architecture of the 8085 matches that detailed in Fig. 1.2, except that the data bus and half of the address bus are multiplexed as they feed to the shared interconnecting pins. The register set is illustrated in Fig.

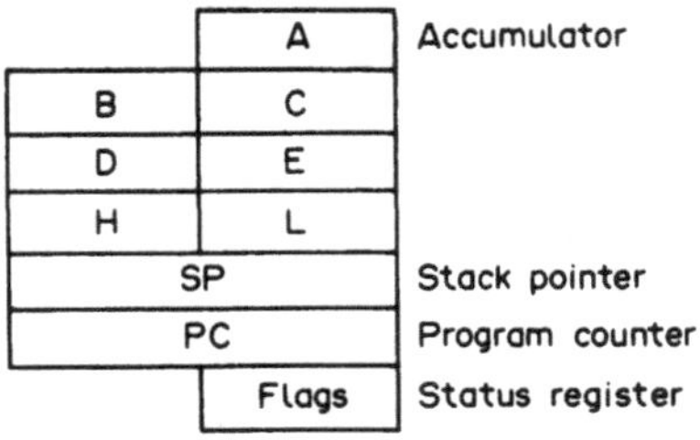

FIG. 2.2. Register set for 8080 CPU.

2.2. In addition to the Accumulator there are six further 8-bit work registers (B, C, D, E, H and L). These can be combined into 16-bit register-pairs (BC, DE and HL) for 16-bit working in several instructions. The HL register-pair is used for register indirect addressing, e.g.

MOV A,M

moves the data byte held in the memory location, which is stored in the HL register-pair, into A register (Accumulator).

The 8085 possesses the following five external interrupts, which are listed together with their interrupt service routine (ISR) start addresses:

RST5.5—start address hex. 002C
RST6.5—start address hex. 0034
RST7.5—start address hex. 003C
TRAP —start address hex. 0024
INTR —start address generated by interrupting device (e.g. a 3-byte CALL instruction).

The program counter is forced to 0 on RESET.

The status register contains five flag bits: Zero, Parity, Carry and Auxiliary Carry.

2.4 Instruction Set

The Intel 8085 instruction set is listed in Table 2.1. The choice of mnemonics (normally 3 or 4 letter names given to instructions) is particularly meaningful in the Intel instruction set. Notice that the legend "r" is used to denote any register (A, B, C, D, E, H or L), whilst the suffix "X" at the end of an instruction is used to denote that a 16-bit register-pair is used in the instruction.

TABLE 2.1 *Intel 8085 Instruction Set*

Mnemonic	Description
Data move	
MOV r1,r2	Move contents of register 2 to register 1
MOV M,r	Move contents of register to memory (using HL)
MOV r,M	Move contents of memory to register
MVI r,data	Load register immediate
MVI M, data	Load memory immediate
LXI rr,data	Load register-pair immediate (rr = B, D, H or SP)
STAX rr	Store A indirect (using register-pair)
LDAX rr	Load A indirect (using register-pair)
STA address	Store A in memory
LDA address	Load A from memory
SHLD address	Store H & L
LHLD address	Load H & L
XCHG	Exchange D & E and H & L registers
IN address	Input to A
OUT address	Output from A
PUSH r	Push register to stack
POP r	Pop register from stack
Data modify	
INR r	Increment register
DCR r	Decrement register
INR M	Increment memory
DCR M	Decrement memory
INX rr	Increment register-pair
DCX rr	Decrement register-pair
ADD r	Add register to A
ADC r	Add register to A with carry
ADD M	Add memory to A
ADC M	Add memory to A with carry
ADI data	Add immediate to A
ACI data	Add immediate to A with carry
DAD rr	Add register-pair to HL
SUB r	Subtract register from A
SBB r	Subtract register from A with borrow
SUB M	Subtract memory from A
SBB M	Subtract memory from A with borrow
SUI data	Subtract immediate from A
SBI data	Subtract immediate from A with borrow
ANA source	AND source (register or memory) with A
XRA source	EX OR source with A
ORA source	OR source with A
CMP source	Compare source with A
ANI data	AND immediate with A
XRI data	EX OR immediate with A
ORI data	OR immediate with A
CPI data	Compare immediate with A
RLC	Rotate A left
RRC	Rotate A right
RAL	Rotate A left through carry
RAR	Rotate A right through carry

continued

TABLE 2.1 *Continued*

Mnemonic	Description
CMA	Complement A
DAA	Decimal adjust A
STC	Set carry flag
CMC	Complement carry flag
Jump	
JMP address	Jump unconditional
Jcc address	Jump on condition cc (signified by status bit)
CALL address	Call subroutine
Ccc address	Call subroutine on condition cc
RET	Return
Rcc	Return on condition cc
Control/miscellaneous	
EI	Enable interrupts
DI	Disable interrupts
RIM	Read interrupt mask
SIM	Set interrupt mask
NOP	No-operation
HLT	Halt
RST	Restart

The addressing modes available, with examples of their use in instructions, are as follows:

(a) Register
```
MOV A,D      ;Move contents of D register to A register (accumulator)
```
(b) Memory (Intel call "direct")
```
LDA 30B6H    ;Load A register with contents of memory location hex. 30B6
```
(c) Register indirect
```
MOV A,M      ;Load A register with contents of memory location that is held in HL register-pair
```
(d) Immediate
```
MVI A,7      ;Load A register with 7 (8-bit number)
```

The addressing mode with JUMP and CALL instructions is (b) above, i.e. the memory address that is to be placed in the program counter when the instruction is completed is held in the second and third bytes of the instruction. Notice that the IN (input) and OUT (output) instructions use only a single-byte port address as an operand, and are therefore two bytes long.

The following program examples illustrate the use of the instruction set:

Program Example 1

```
                                                                   Bytes
      MVI A,0       ;Clear A                                         2
      MVI E,6       ;Load loop count of 6 into E                     2
      LXI H,4000H   ;Load HL with memory address 4000                3
LOOP: ADD M         ;Add from memory to A (using HL)                 1
      INX H         ;Increment HL                                    1
      DCR E         ;Decrement loop count in E                       1
      JNZ LOOP      ;Jump around loop (6 times)                      3
      HLT           ;Halt                                            1
```

The program adds 6 numbers held at memory addresses hex. 4000 through 4005 into A register (the accumulator). Notice that the number of bytes required for each instruction is shown on the right-hand side. The program commences by loading A, E and HL registers using immediate addressing mode instructions. It then enters a loop, and the first instruction is given a label of "LOOP" to represent the memory address at which it is stored. During each pass of the loop a different number is added into A register, the HL register-pair is incremented and the loop count in E is decremented. On the sixth pass through the loop E is decremented to zero, and the program halts.

Program Example 2

```
        IN  30H      ;Input to A from address 30
        CPI 1        ;Compare A with binary 0000 0001
        JZ  FINISH   ;Jump if zero (right-hand bit set to 0)
        STA 0600H    ;Store input byte in memory
        MVI A,4      ;Load A with 0000 0100
        OUT 31H      ;Output A to address 31
FINISH: JMP 8000H    ;Jump to new program
```

The program illustrates the use of input/output instructions, which utilise single-byte operands, i.e. input/output addresses are only 8 bits long. The program reads in an input byte (perhaps from a block of 8 pushbuttons). Bit 0 (the right-hand bit) is examined, and if it is set to 0 a program jump is obeyed to the label FINISH. Program control is then transferred to another program, which commences at memory location 8000. If the bit tested is set to 1, e.g. a pushbutton is pressed, the input byte is stored in memory and a logic 1 is sent to bit 2 on an output address, perhaps to illuminate an indicator LED.

2.5 Support Devices

Intel support the 8080 and 8085 CPUs with the widest range of devices offered by 8-bit microprocessor manufacturers. They offer four PIOs which include variable quantities of memory (ROM and RAM) as follows:

(a) 8155—3 ports, 256 bytes RAM, 14-bit counter
(b) 8255—3 ports
(c) 8355—2 ports, 2K ROM
(d) 8755—2 ports, 2K EPROM

The pin functions of the ubiquitous 8155 are shown in Fig. 2.3. Ports A and B each provide 8 input or output signal lines, but Port C is only 6-bits wide. Each port direction is programmable.

The 8 address lines that are multiplexed with the data lines can be used to address one of the 256 RAM bytes. Alternatively, if an input/output instruction is executed in place of a memory transfer instruction, the address lines select one of the following addresses on the device:

A2	A1	A0
0	0	0—Control register
0	0	1—Port A
0	1	0—Port B
0	1	1—Port C
1	0	0—Counter/timer (low order 8 bits)
1	0	1—Counter/timer (high order 6 bits)

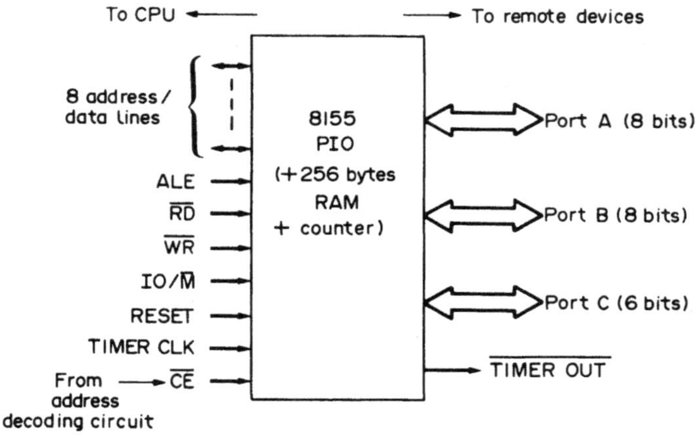

FIG. 2.3. 8155 PIO circuit connections.

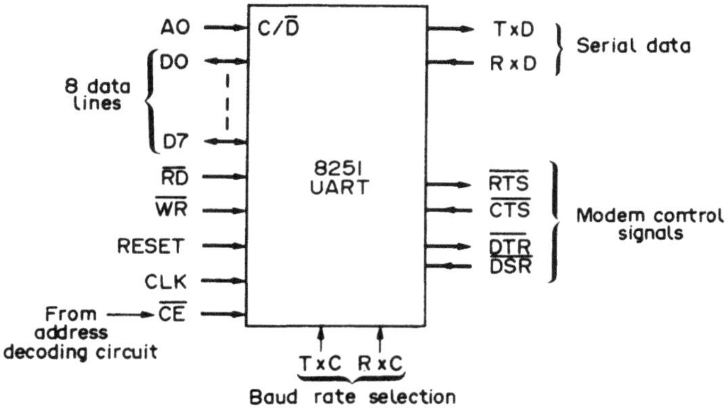

FIG. 2.4. 8251 UART circuit connections.

The control register is used to set port directions, e.g.

bit 0 = Port A direction (0 = input, 1 = output)
bit 1 = Port B direction (0 = input, 1 = output)
bits 2 and 3 = Port C direction (00 = input, 11 = output, 01/10 = mixed)
bits 6 and 7 = Timer control (11 = start timer immediately)

The counter can be used to count external pulses, or (if fed from the CPU clock) to produce a time delay, with a typical maximum delay of:

hex 3FFF (maximum 14-bit count) × 0.5 μs (2 MHz clock)
= 8 ms (approx.)

Alternatively the TIMER OUT signal can be connected to a CPU interrupt line to trigger an interrupt service routine at regular intervals. A further application is as a clock signal to a UART.

The standard Intel UART that interfaces an 8080/8085 system to a serial-drive peripheral, e.g. VDU or data link, is the 8251. The circuit connections are shown in Fig. 2.4. The device is universal, i.e. it is designed to connect to almost any microprocessor. Therefore data bus and address bus lines are not multiplexed. The single address line A0 must be de-multiplexed from AD0 by a simple external circuit when the device is connected to an 8085 CPU. It is connected to the C/D (Control/Data) signal pin on the 8251, and the two states of this signal combine with the setting of either RD (Read) or WR (Write) to produce four addressable registers on the 8251, as follows:

$C/\overline{D}$ (A0)	$\overline{RD}$	$\overline{WR}$
0	1	0—Transmit data
0	0	1—Receive data
1	1	0—Control register
1	0	1—Status register

The control register is used to select the number of data bits, number of stop bits, parity and baud rate factor. The latter parameter selects the frequency divider ratio (1, 16 or 64) that is applied to the incoming transmitter and receiver clock signals (TxC and RxC). To obtain the full range of baud rates the clock signals must be variable and programmable, e.g. generated by a counter on a 8155 PIO described above.

Intel supply the 8257 DMA controller, which provides for four DMA channels. Two input/output devices that commonly use DMA channels are:

(a) 8271 floppy disc controller (FDC)
(b) 8275 CRT controller.

Figure 2.5 is a simplified representation of an input/output arrangement that

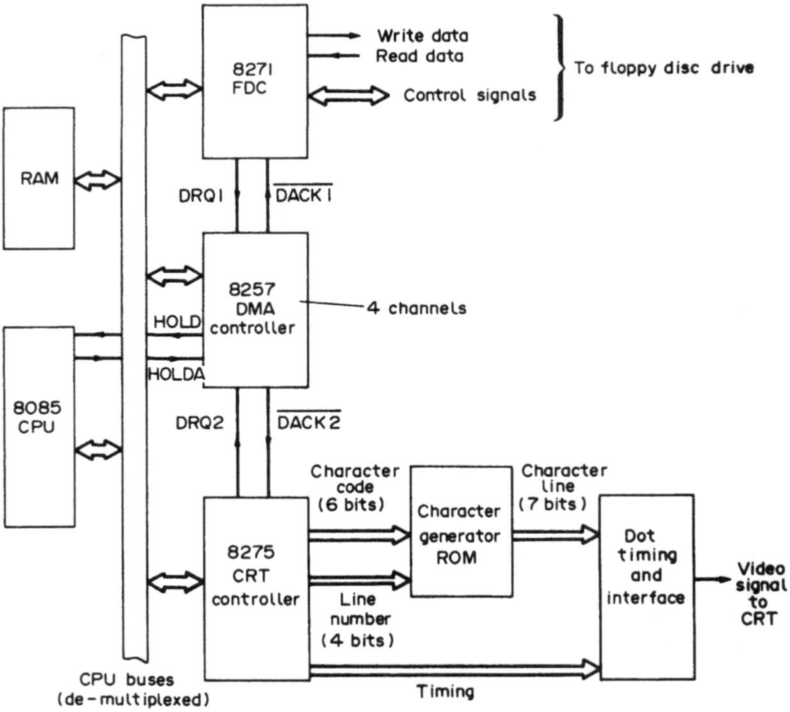

FIG. 2.5. Intel DMA control system (for FDC and CRT controller).

utilises these three devices. The actions required to transfer block of bytes between memory (RAM) and floppy disc are as follows:

(1) Software initiates a transfer by loading the DMA controller with the start address of memory, and the number of bytes to be transferred. Additionally software loads the FDC with the disc address (track and sector).
(2) FDC requests DMA transfers (DRQ1) when the required disc sector is beneath the read/write head. The DMA controller requests the use of the system buses from the CPU (using HOLD), which responds with HLDA.
(3) Bytes are transferred directly between memory and floppy disc—DMA controller generates each memory address, and FDC processes each data byte. DMA transfers can be continuous ("burst").

The operation of the CRT controller circuit, which extracts character bytes from main memory and creates a video signal to drive a CRT display, is as follows:

(1) The DMA controller is initiated by software to hold the start address of video data in main memory RAM, and the number of bytes to be transferred.
(2) System software places the required character bytes into RAM (typically a reserved 2K bytes area for 80 column, 24 line CRT display).
(3) CRT controller requests DMA transfers (DRQ2) repeatedly. The DMA controller responds with $\overline{\text{DACK2}}$ and acquires use of the buses from the CPU.
(4) Typically 80 character bytes are transferred from main memory RAM to be stored in CRT controller—the DMA controller generates each memory address, and the CRT controller receives each data byte.
(5) CRT controller presents each of 80 characters in turn to Character Generator ROM which holds dot matrix for each character code—6-bit character code combines with 4-bit line number to address ROM to produce 7-bit dot pattern for each line of character. This is repeated for each line in the dot matrix for entire block of 80 characters.
(6) Whilst one of 24 rows of 80 characters is being processed as in (5) the next row of 80 characters is transferred by DMA to the CRT controller.
(7) Process is repeated—from stage (3) onwards—for each row of 80 characters until a complete CRT screen is displayed. This screen refresh is repeated continuously such that the latest information held in video RAM is displayed on the CRT.

Additional devices offered by Intel include:

(a) 8231 arithmetic processor (add, subtract, multiply, divide, trigonometric and logarithmic functions)
(b) 8232 floating point (32- or 64-bit numbers) processor
(c) 8202 dynamic RAM controller
(d) 8253 CTC (three 16-bit counters).

2.6 Applications

Figure 2.6 illustrates a simple circuit, comprising only three ICs, which offers a complete microcomputer system. The circuit package comprises:

(a) 8085 CPU
(b) 256 bytes RAM
(c) 2K bytes EPROM
(d) 5 ports (one is only 6 bits wide)—total of 38 bits
(e) 1 counter/timer.

Such a configuration is suitable for a wide range of industrial applications, e.g. sequence control system involving the input of several contact-closure signals and the output of several on/off control signals.

A generalised representation of a cash register control board is shown in Fig. 2.7. Two 8155 PIOs are applied to interface to the printer, segment display unit and keyboard. A total of 512 bytes of RAM is available for use by the control program, which is held in a 4K ROM. An RS232-C link is provided to connect the cash register to a remote stock control computer which updates its stock records after every customer transaction.

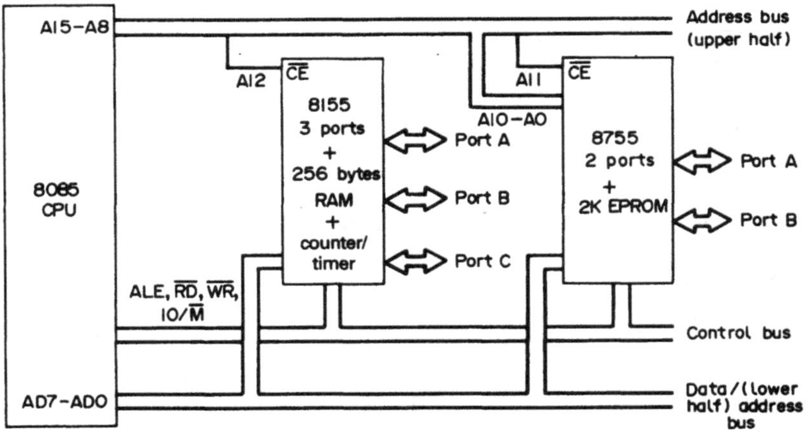

FIG. 2.6. Intel 8085 microcomputer system (general application).

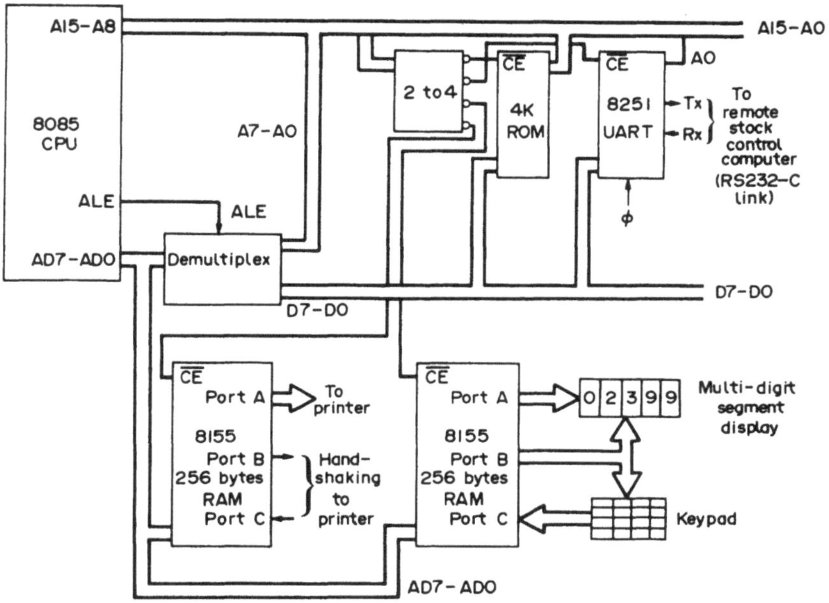

FIG. 2.7. Intel 8085 cash register configuration.

2.7 Single-chip Microcomputers

Intel manufacture a range of 8-bit single-chip microcomputers that supply a range of support circuits (memory and input/output) on-chip with the CPU. Although the CPU function is 8-bit for each device, none of the family is strictly based on the 8080/8085 instruction set. The family includes:

(a) 8048—3 ports, 1 counter, 64 bytes RAM, 1K ROM
(b) 8748—3 ports, 1 counter, 64 bytes RAM, 1K EPROM
(c) 8035—3 ports, 1 counter, 64 bytes RAM (no ROM)
(d) 8049—3 ports, 1 counter, 128 bytes RAM, 2K ROM
(e) 8022—3 ports, 1 counter, 64 bytes RAM, 1K ROM, 1 A/D converter
(f) 8051—3 ports, 1 counter, 128 bytes RAM, 4K ROM, 1 UART, extra instructions (e.g. multiply and divide).

The pin functions of the popular 8748 are illustrated in Fig. 2.8. There is a single interrupt line to augment the $\overline{\text{RESET}}$ signal. Although the chip is designed to operate as a complete functional system, it is possible to access external Intel support devices, e.g. 8155 PIO, to provide additional facilities.

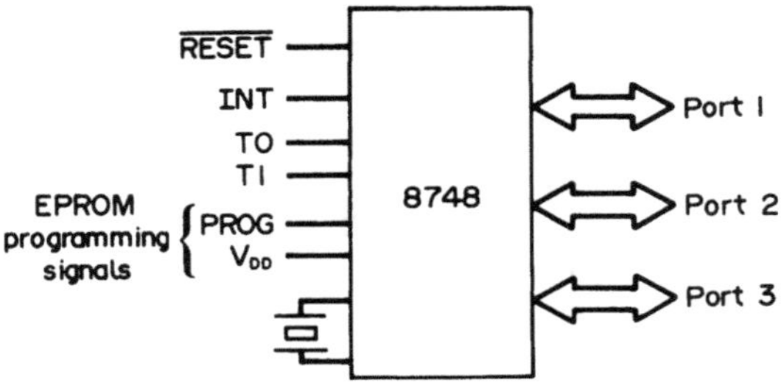

FIG. 2.8. Pin functions of 8748 single-chip microcomputer.

Single-chip microcomputers (sometimes called "microcontrollers") are particularly suited to single-function domestic controllers, e.g. washing machine controllers, telephone answering machines.

Bibliography

1. *Microprocessors and Microcomputers.* Barry G. Woollard. McGraw-Hill, 1981.
2. *Microprocessors and Programmed Logic.* Kenneth L. Short. Prentice-Hall, 1981.
3. *Study Notes for Technicians: Microelectronic Systems Levels 1 & 2.* R. C. Holland. McGraw-Hill, 1983.

CHAPTER 3

ZILOG Z80

3.1 Introduction

The design of the Zilog Z80 was based on the Intel 8080/8085, but it possesses several additional powerful facilities. It has achieved a diverse range of applications in office computer systems, home computers and industrial controllers.

The Z80 is "upwards compatible" with the Intel 8085 in terms of instruction set and machine code, i.e. programs written for the 8085 will run on a Z80. However, compatibility in the other direction is not always achieved because of the additional instructions which the Z80 possesses.

The Z80A is designed to operate at 4 MHz, whilst the Z80B will run at up to 6 MHz.

3.2 Z80 Pin Functions

Figure 3.1 illustrates the pin functions of the Z80. The data bus is not multiplexed with one-half of the address bus on the same pins, as applied on the Intel 8085. Consequently there are fewer pins available for control bus lines, and only three interrupts are used: $\overline{\text{RESET}}$, $\overline{\text{NMI}}$ (Non Maskable Interrupt and $\overline{\text{INT}}$. However, this is not normally a limitation because the last signal can be shared by a large number of Zilog interrupting devices (described in section 3.5).

The Z80 does not possess an on-chip clock circuit, and the clock generator must be supplied externally.

Most control signals are self-explanatory, e.g. the DMA control signals are $\overline{\text{BUSRQ}}$ (Bus Request) and $\overline{\text{BUSAK}}$ (Bus Acknowledge), but the $\overline{\text{RFSH}}$ (Refresh) signal is unique to the Zilog CPUs. It is set by the Z80 between each instruction and is designed to assist in refeshing any dynamic RAM that may be connected to the CPU buses.

3.3 Processor Architecture

The register set of the Z80 CPU is shown in Fig. 3.2. The work registers A, B, C, D, E, H and L are identical to those applied in the Intel 8085, but the Z80 possesses the unusual feature of offering a duplicate set of registers

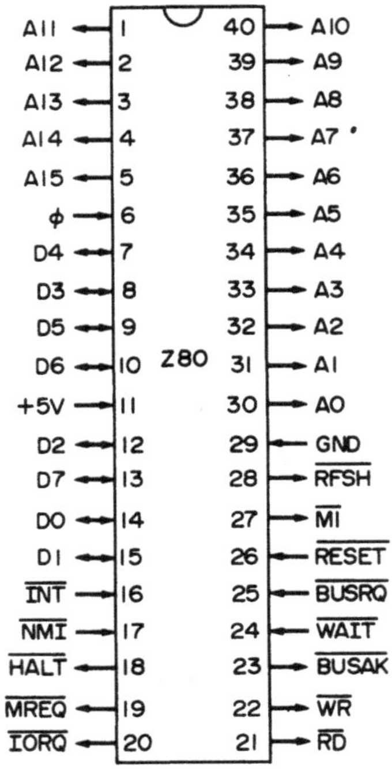

FIG. 3.1. Z80 pin functions.

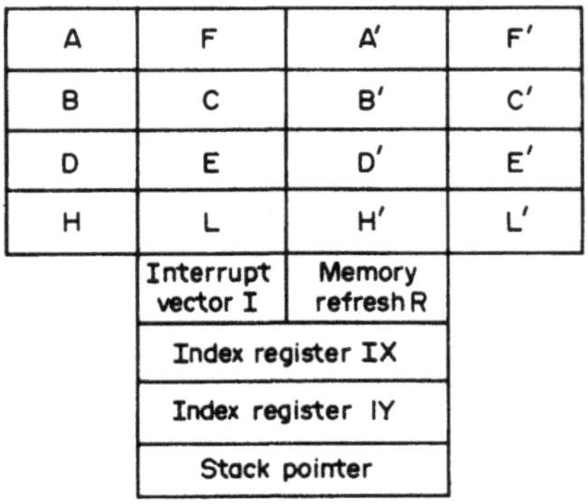

FIG. 3.2. Register set for Z80 CPU.

(A′ to L′) and status register (F′). This feature is useful when a subroutine or interrupt service routine (ISR) is entered, because the programmer can select the alternative set within the subroutine or ISR, thus avoiding the necessity of storing away the contents of the main program's registers, e.g. with PUSH instructions. The following two instructions implement the changeover (in either direction):

```
EXX          ;Exchange BC, DE and HL
EX  AF,AF'   ;Exchange AF (F is Flags)
```

The interrupt vector I is an 8-bit register that is employed by the CPU to locate the start address of an ISR when the CPU is selected to operate in one of three "interrupt modes" (modes 0, 1 and 2). The start addresses of ISRs can be summarised as follows:

$\overline{\text{RESET}}$—PC is set to 0000 (PC = program counter)
$\overline{\text{NMI}}$ —PC is set to 0066
$\overline{\text{INT}}$ —CPU obeys contents of interrupt vector I for CPU "mode 0"
—PC is set to 0038 for CPU "mode 1"
—CPU examines location XXYY to find start address of ISR for CPU "mode 2" (XX is contents of interrupt vector I, YY is supplied on data bus from interrupting device).

A program instruction (e.g. IM 2) is required to set the CPU into either interrupt modes 0, 1 or 2, and if mode 2 is selected the interrupting device (e.g. PIO,CTC) must be initialised with the 8-bit value (YY) which it presents back to the CPU when it generates an interrupt.

The memory refresh register R is applied together with the $\overline{\text{RFSH}}$ signal (see section 3.2) to refresh dynamic RAM. The register is incremented after each instruction and is placed on the lower half of the address bus between instructions.

The two 16-bit index registers (IX and IY) allow indexed addressing mode instructions to be used.

3.4 Instruction Set

The Z80 instruction set is listed in Table 3.1. Although identical machine code for Intel 8085 is retained by the Z80, instruction mnemonics are different. The Z80 has several additional instructions to the 8085, e.g. data block move instructions and individual bit checking/setting instructions. The addressing modes available are:

(a) Register

```
LD A,C              ;Load A with contents of C register
```

(b) Memory

```
LD A,(5020H)    ;Load A with contents of memory location 5020
```

(c) Indirect

```
LD A,(HL)       ;Load A with contents of memory location held in
                 HL register-pair
```

(d) Immediate

```
LD A,3          ;Load A with 03 (8-bit number)
```

(e) Indexed

```
LD A,(IX+2)     ;Load A with contents of memory location held in
                 IX, with 2 (8-bit "displacement") added to that
                 address
```

There are two addressing modes available with Jump instructions, as follows:

(f) Absolute jump

```
JP NZ,0500H     ;Jump (if not zero) to address 0500
```

(g) Relative jump

```
JR  Z,-9        ;Jump (if zero) relative −9 memory locations (i.e.
                 9 bytes back through memory from address of
                 next instruction)
```

TABLE 3.1 *Z80 Instruction Set*

Mnemonic		Description
Data move		
LD	r1,r2	Move contents of register 2 to register 1
LD	(addr),A	Store A into memory
LD	A,(addr)	Load A from memory
LD	(rr),A	Store A indirect (using register-pair)
LD	A,(rr)	Load A indirect (using register-pair)
LD	r,n	Load register immediate (or register-pair)
LD	r,(IX+d)	Load register indirect (IX or IY) with displacement d
LD	(IX+d),r	Store register indirect (IX or IY) with displacement d
LD	SP,rr	Load stack pointer from register (HL, IX or IY)
LDD		Load location (DE) with location (HL), decrement DE, HL and BC
LDDR		Load location (DE) wih location (HL), decrement DE, HL and BC, repeat until BC=0
LDI LDIR		As above, but increment DE, HL and BC
EX	DE,HL	Exchange DE and HL
EXX		Exchange BC, DE and HL
EX	AF,AF'	Exchange A and F
EX	(SP),rr	Exchange location (SP) and register-pair
In	A,(n)	Input to A from input/output address n
OUT	n,(A)	Output from A to input/output address n
plus additional block move instructions to input/output addresses		

TABLE 3.1 *Continued*

Mnemonic		Description
Data modify		
INC	source	Increment (various addressing modes)
DEC	source	Decrement (various addressing modes)
ADC	A,source	Add to A with carry
ADC	HL,rr	Add to HL register-pair with carry
ADD	A,source	Add to A
SBC	A,r	Subtract register from A with carry
SBC	HL,rr	Subtract register-pair from HL with carry
SUB	r	Subtract register from A
AND	source	AND source with A
OR	source	OR source with A
XOR	source	Exclusive-OR source with A
BIT	b,source	Test bit b of source
CP	source	Compare source with A
plus several multi-function Compare instructions		
RES	b,source	Reset bit b of source
RL	r	Rotate left through carry
RLA		Rotate left A through carry
RLC	source	Rotate left circular
RR	r	Rotate right through carry
RRA		Rotate right A through carry
RRC	source	Rotate right circular
SET	b,source	Set bit b of source
SLA	r	Shift left arithmetic
SRA	r	Shift right arithmetic
SRL	r	Shift right logical
CPL		Complement A
DAA		Decimal adjust A
NEG		Negate A
CCF		Complement carry flag
SCF		Set carry flag
Jump		
JP	addr	Jump unconditional
JP	cc,addr	Jump conditional (cc refers to status bits)
JR	d	Jump relative unconditional (d = displacement)
JR	cc,d	Jump relative conditional
CALL	addr	Call subroutine
CALL	cc,addr	Call subroutine on condition cc
RET		Return
RET	cc	Return on condition cc
DJNZ	e	Decrement B and jump relative if B is not zero
Control/miscellaneous		
DI		Disable interrupts
EI		Enable interrupts
HALT		Halt
IM	n	Set interrupt mode n (n = 0, 1 or 2)
NOP		No operation
RST	addr	Restart to memory address

The following program examples illustrate use of the instruction set.

Program Example 1

```
        LD   A,0FFH     ;Load A with FF (all 1s)
        LD   D,10       ;Load loop count of 10 into D
        LD   BC,3000H   ;Load BC with 3000
CONT:LD   (BC),A     ;Store A in memory (using BC indirect)
        INC  BC         ;Increment BC
        DEC  D          ;Decrement loop count in D
        JP   NZ,CONT    ;Jump around loop (10 times)
        HALT            ;Halt
```

This program stores FF in 10 memory locations commencing at address 3000.

Program Example 2

```
;7E is address of 8-bit port to carry data bits (in ASCII)
;7F is address of handshaking signal "out" bit (strobe), and also
;of handshaking "in" bit (ready).

        LD    D,25       ;Set loop count of 25
        LD    BC,0700H   ;Set start address of 25 ASCII characters
  BUSY:IN    A,(7FH)    ;Input ready signal
        BIT   0,A        ;Check bit 0 in A (logic 0 if printer is busy)
        JP    Z,BUSY     ;Poll "ready" until it indicates printer ready
        LD    A,(BC)     Load A with ASCII character
        OUT   (7EH),A    ;Output to port (8 data bits)
        LD    A,0        ;
        OUT   (7FH),A    ;Output 0 to "strobe" signal
        CALL  DELAY      ;Delay
        LD    A,1        ;
        OUT   (7FH),A    ;Output 1 to "strobe" signal
        CALL  DELAY      ;Delay
                         ;A complete pulse has now been set on
                          "strobe"
        INC   BC         ;Step on to next character
        DEC   D          ;Decrement loop count
        JP    NZ,BUSY    ;Repeat loop (25 times)
        JP    0          ;Jump to other program (at address 0000)
;End of program, now for subroutine
DELAY:LD    E,0FFH     ;Set delay loop count
 LOOP:NOP
        NOP
        NOP
        DEC   E          ;Decrement loop count
        JP    NZ,LOOP    ;Repeat loop
        RET
```

This program outputs a string of characters to a Centronics interface parallel printer. Within the loop that outputs characters it firstly polls the "ready" bit (handshaking response bit from printer) to ensure that the printer is not already printing a character. Next the character is sent to the printer. Finally a pulse is applied on the "strobe" bit (handshaking bit to printer), which causes the printer to examine the 8 data/character bits and print the character. The loop is repeated 25 times.

3.5 Support Devices

Zilog supply a PIO, CTC and dual-UART (called DART) to support the Z80 CPU. If required these devices can be set to operate in interrupt mode (e.g. the PIO can interrupt when a specific input line is set, the CTC can interrupt when a counter reaches zero, the DART can interrupt when a character is received). Because the Z80 possesses only two interrupt lines (apart from $\overline{\text{RESET}}$), it is customary to arrange that these devices share the same interrupt line $\overline{\text{INT}}$ in a "daisy-chain" system illustrated in Fig. 3.3. Any one of the devices can generate an interrupt, but the daisy-chain ensures a priority system, with the PIO having the highest priority and the DART having the lowest priority in this arrangement. Logic 1 is fed to the IEI (Interrupt Enable In) pin on the PIO, whence the signal is "chained" through each other device. If the CTC generates an interrupt, the daisy-chain is broken (IEO is set to logic 0), such that the DART cannot generate an interrupt. When the interrupt service routine for the CTC is completed a RETI (Return from Interrrupt) instruction is executed, and this opcode is detected on the data bus by the CTC, which re-establishes the daisy-chain. The CPU can distinguish which device is generating the interrupt, because

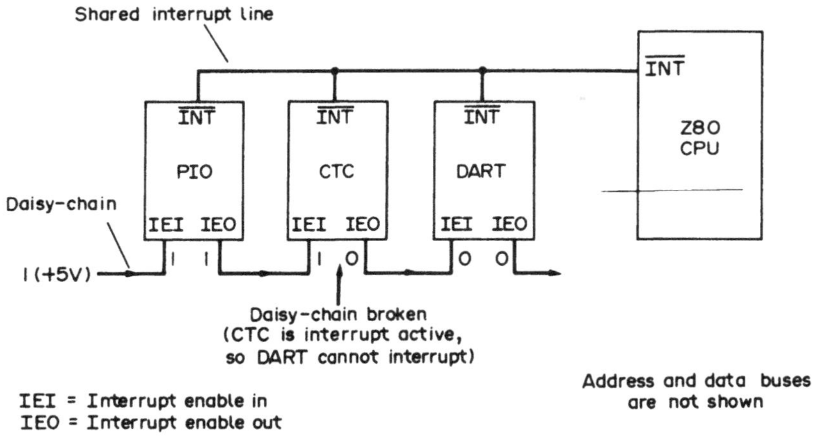

FIG. 3.3. Z80 interrupt "daisy-chain".

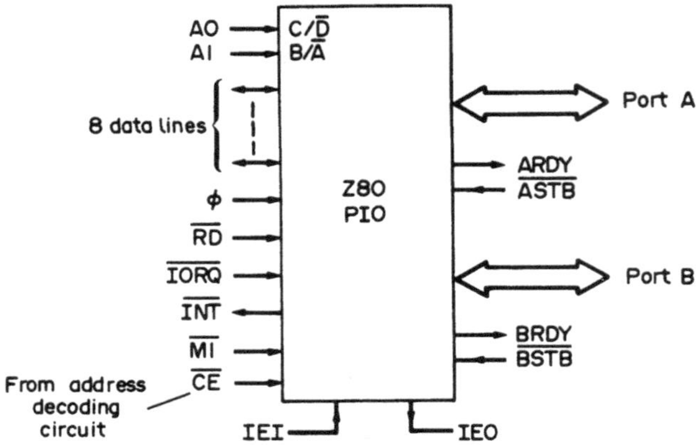

FIG. 3.4. Z80 PIO circuit connections.

the interrupting device places an 8-bit "code" on the data bus—see $\overline{\text{INT}}$ (CPU mode 2, in section 3.3).

The pin functions of the popular Z80 PIO are illustrated in Fig. 3.4. The device possesses two 8-bit ports, with handshaking signals available for each port. The four addresses on the device are:

B/$\overline{\text{A}}$	C/$\overline{\text{D}}$
0	0—Port A data
0	1—Port A control
1	0—Port B data
1	1—Port B control

Therefore each port possesses its own control register. This enables each port to have a wide range of programmable options, and these include byte output, byte input and mixed mode (some bits output, some bits input). Port A alone can be selected to be bi-directional. The control register for each port is used additionally when the 8-bit interrupt code is loaded into the device.

The Z80 CTC is illustrated in Fig. 3.5. The device possesses four 8-bit counters. Each counter (or "channel") can be set to decrement by a clock pulse (ϕ) or by an external pulse (CLK/TRn). The control register that is used to select such options possesses the same address as the counter itself (bit 0 in the data byte sent to the device must be set to 1 to indicate that control mode is selected). Similarly the interrupt mode, if required, is programmed into the device using the same channel address. Typical applications of the device are:

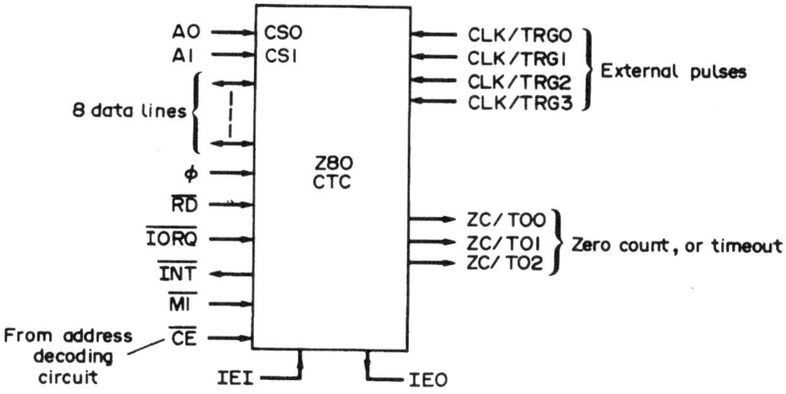

FIG. 3.5. Z80 CTC circuit connections.

(a) decrement a count on each clock pulse in order to generate an interrupt at regular intervals, e.g. every 1 ms to drive a real-time clock
(b) generate a stream of pulses on one of the ZC/T0n outputs to provide a clock signal to a UART
(c) count external pulses (entered on one of the CLK/TRGn inputs)

The Zilog UART is named a DART (Dual-UART) and is illustrated in Fig. 3.6. It offers two completely separate RS232-C channels, each with its own control register for initialising baud rate, parity, etc. The Zilog SIO (Serial Input Output) is almost identical in operation but offers the additional facility of synchronous transmission.

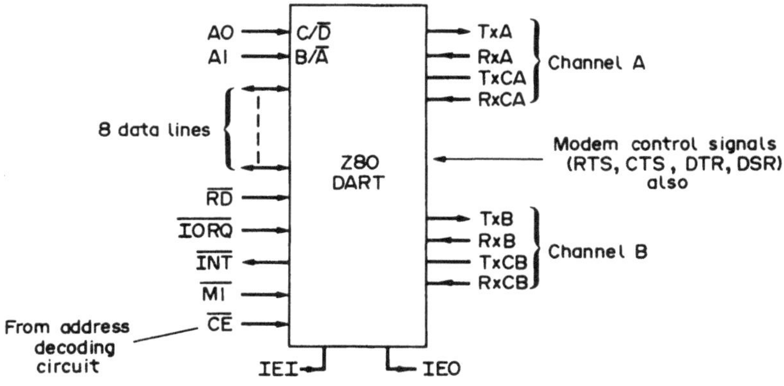

FIG. 3.6. Z80 DART circuit connections.

3.6 Applications

The Z80 is used in a large number of home computers (e.g. Sinclair ZX81, Sinclair Spectrum, Tandy TRS80) and office computers (e.g. Sharp MZ80B, Research Machines RML380Z, Icarus Superbrain). A simplified representation of the Sinclair ZX81 is shown in Fig. 3.7. This home computer possesses only a small amount of memory and no direct peripheral interfacing. The circuit functions of address decoder, video generation and keyboard scanning are performed by a single custom-designed IC (called "Sinclair Computer Logic" in the diagram).

A larger computer configuration that is applied in an industrial environment for control of food packing systems is shown in Fig. 3.8. The computer system monitors the weights of the contents of 16 hoppers, which each contain quantities of food products. The hoppers are designed to hold one-quarter of the total weight to be dispensed into a packet. The control program must therefore evaluate which 4 of the 16 hoppers must be selected to feed into the packet to produce the total weight that is closest to the required weight. The computer opens the required hopper gates. The control program is held in three 4K ROM chips, and a 2K RAM is used for the storage of temporary data and for the system stack. Two 8255 PIOs (see chapter 2) provide plant interfacing. One PIO connects to an A/D device which additionally provides analogue selection (a 4-bit code switches 1 of 16 analogue signals to the 10-bit A/D circuit). The second PIO outputs single-bit signals to solenoids which open the hopper outlet gates.

A much-simplified representation of a multi-board office computer (Comcen Crystal) that is based on a Z80 is shown in Fig. 3.9. This supports several peripherals—VDU (RS232-C link), printer (Centronics interface) and floppy disc (dual-drive).

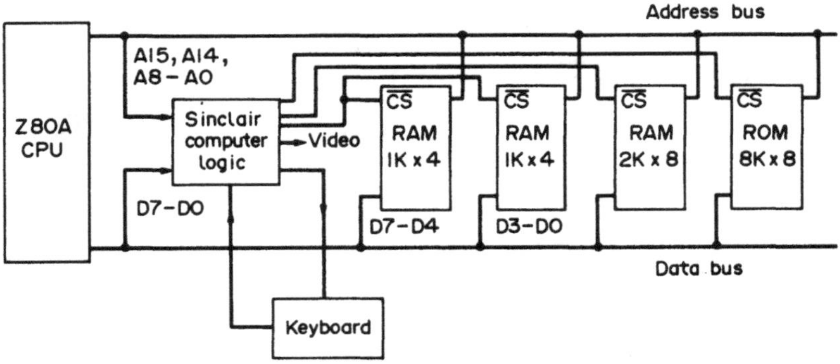

FIG. 3.7. Simplified schematic of Sinclair ZX81.

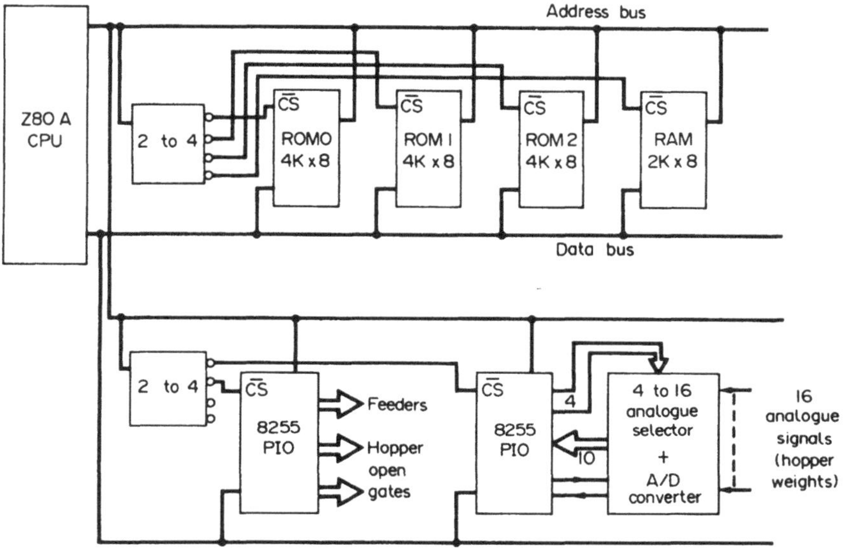

FIG. 3.8. Simplified schematic of weighing hopper control system.

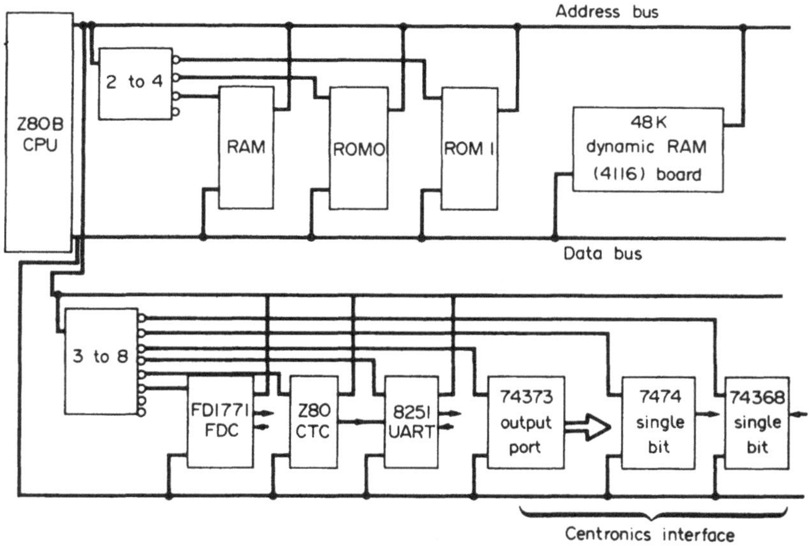

FIG. 3.9. Simplified schematic of Comcen Crystal office computer.

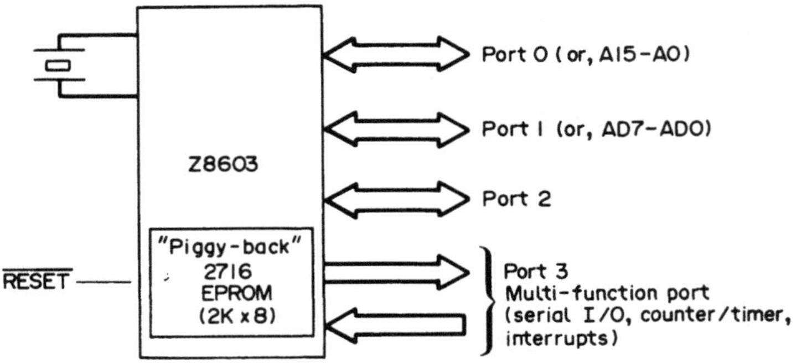

FIG. 3.10. Zilog Z8603 pin functions.

3.7 Single-chip Microcomputers

The Zilog Z8 family of single-chip microcomputers offers a range of fast and powerful processors. A typical member of the family offers 2K (or 4K) bytes of ROM/EPROM, 144 bytes of RAM, a UART, 2 counter/timers and 4 ports (32 input/output signals). The family comprises:

(a) Z8601—2K ROM plus features above
(b) Z8602—ROMless, 64-pin package with memory bus
(c) Z8603—2K ROM plus features above + 2K EPROM (see Fig. 3.10)
(d) Z8681—ROMless, 40-pin package with memory bus
(e) Z8611—4K version of (a)
(f) Z8612—4K version of (b)
(g) Z8613—4K version of (c)
(h) Z8671—as (a) plus tiny-BASIC interpreter

The Z8 family instruction set is completely different from that of the Z80. Each Z8 processor contains 144 bytes of on-chip register/RAM. These registers include the input/output addresses (e.g. ports, timers and UART) as well as 124 general work registers, which can all operate as accumulators or index registers. Notice from Fig. 3.10 that Ports 0 and 1 can be used to access the address and data buses if it is required to expand memory or input/output externally.

Bibliography

1. *Z80 Microprocessor Programming and Interfacing*, Books 1 and 2. Elizabeth A. Nichols, Joseph C. Nichols and Peter R. Rony. Prentice-Hall, 1979.
2. *Z80 Applications*. James W. Coffron. Sybex, 1983.
3. *Programming the Z80*. Rodney Zaks. Sybex, 1982.

CHAPTER 4

MOTOROLA MC6800 FAMILY (8-BIT)

4.1 Introduction

Motorola developed the MC6800 at the same time as Intel were developing the 8080. Following the initial popularity of this device, Motorola introduced other CPUs into the family, culminating in the MC6809. This latter device was introduced after the other popular 8-bit microprocessors offered by rival manufacturers, and is probably the most powerful 8-bit device, with a wide range of enhanced features.

4.2 MC6800 Pin Functions

Figure 4.1 shows the pin functions of the 40-pin MC6800. The device requires two clock signals ϕ_1 and ϕ_2, which are in antiphase with each other. The MC6800 instructions require between two and eight machine cycles to execute. There are two interrupt signals ($\overline{IRQ}$ and $\overline{NMI}$) in addition to the normal $\overline{RESET}$ signal. When the $\overline{HALT}$ signal is set low, the CPU ceases operation and floats the address and data buses (and R/W). This is useful when it is required to perform program debugging and to execute one instruction at a time. In this condition BA (Bus Available) is set.

VMA (Valid Memory Address) is set whenever a valid address is set on the address bus. DMA requests can be activated by setting TSC (Three State Control) high, in order to float the address bus and R/W, and by setting DBE (Data Bus Enable) high, in order to float the data bus.

Although different pin numbers are applied, the MC6809 has similar pin functions. Notice that two pins are unused on the MC6800, and this allows the MC6809 to offer an additional interrupt signal.

4.3 Processor Architecture

The register set for the MC6800 is illustrated in Fig. 4.2. There are two accumulators, such that the results of ALU operations can be placed in either accumulator within an instruction. This is an advantage over the Intel 8085 and the Zilog Z80, but the larger number of registers in these latter devices outweighs this feature for most programming applications. MC6800

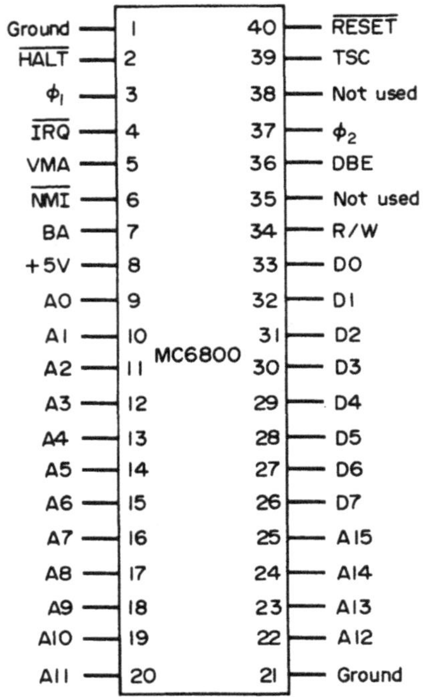

FIG. 4.1. MC6800 pin functions.

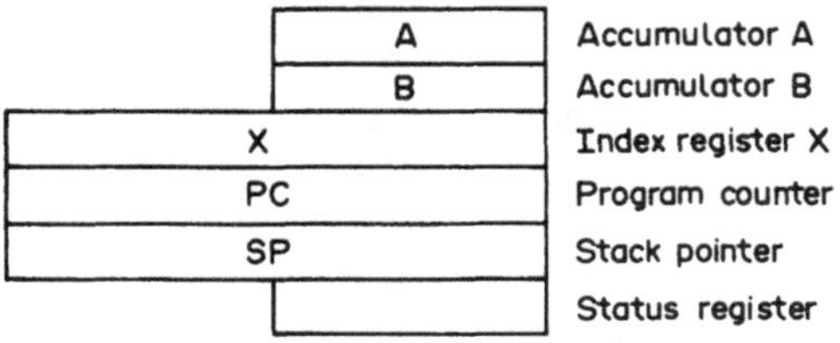

FIG. 4.2. Register set for MC6800 CPU.

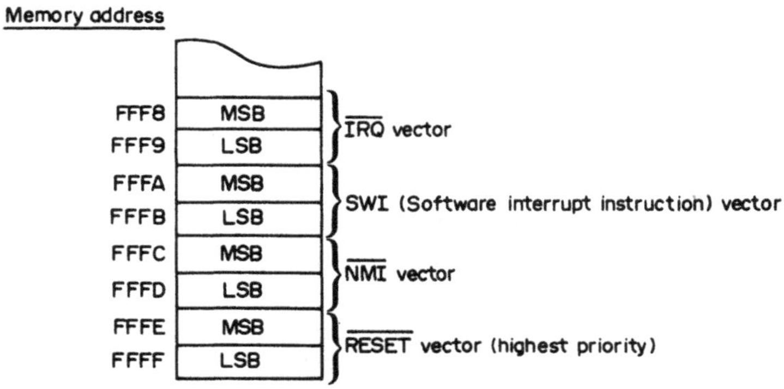

FIG. 4.3. MC6800 inerrupt vectors.

programs are forced to employ memory reference instructions, which are slow, in place of register reference instructions.

The 16-bit index register X is used to modify memory addresses when the indexed mode of addressing is used in an instruction, and this register can be loaded, incremented and decremented. The usual program counter and stack pointer are also 16 bits wide. Only 6 bits in the status register (called "condition code register" by Motorola) are actually used—bits 7 and 6 are permanently set to 1.

The interrupt vectors are listed in Fig. 4.3. They use the last 8 memory locations, and each of the 4 memory addresses is stored with the most significant byte (MSB) first. For example, when $\overline{NMI}$ is set the contents of locations FFFC and FFFD are placed in the program counter and program execution continues from that address. The CPU automatically stores the contents of the register set on the stack in the following order:

Program counter	(LSB)	
Program counter	(MSB)	↓
Index register	(LSB)	↓
Index register	(MSB)	↓ Stack pointer decrementing
Accumulator A		↓
Accumulator B		
Status register		

This automatic saving of the CPU registers is an advantage over the Intel and Zilog devices, which require additional instructions (PUSH) to perform the same function. When the interrupt service routine is completed the RTI instruction must be obeyed in order to return control to the interrupted program and to reload the CPU registers from the stack.

The MC6809 has similar internal architecture to the MC6800, but it does offer the following additional facilities:

(a) a second index register Y
(b) a second stack pointer (the CPU uses the "hardware stack pointer" and the programmer uses the "user stack pointer").

4.4 Instruction Set

The MC6800 instruction set includes 72 basic instructions, but the use of several addressing modes by many instructions expands this number to 197 different machine codes. The set is characterised by a large variety of memory transfer instructions and conditional-jump ("branch on condition") instructions. There are no input/output instructions, and so input/output must be memory mapped, i.e. input/output circuits must be connected as if they are memory circuits.

The addressing modes available, together with examples of their use, are:

(a) Implied
```
INX            ;Increment index register
```
(b) Accumulator
```
COM A          ;Complement accumulator A (invert all bits)
```
(c) Immediate
```
LDA A #$57     ;Load accumulator A with hex 57 (# denotes
                immediate addressing)
```
(d) Direct
```
LDA A $3B      ;Load accumulator A with contents of memory
                location hex 3B
```

This is sometimes called "base page addressing"—the base page is the first 256 bytes of memory. Instructions are 2 bytes long, compared with 3 bytes for the following addressing mode.

(e) Extended
```
LDA A $702F    ;Load accumulator A with contents of memory
                location hex 702F
```
(f) Indexed
```
LDA A $40,X    ;Load accumulator A with contents of memory
                location obtained by adding hex 40 to the contents
                of the index register X
```
(g) Relative
```
BEQ -5         ;Branch if equal to zero backwards 5 bytes
```

Notice that full 16-bit addressing is available with subroutine and unconditional jump instructions, e.g.

```
JMP $5000   ;Jump to memory location 5000
JSR $6570   ;Jump to subroutine at memory location 6570
```

Program Example 1

```
      LDX #$40     ;Load X with hex 40
      LDA A #0     ;Clear A
LOOP: ADD A $00,X  ;Add to A from address 0, indexed by X
      INX          ;Increment X
      CPX #$45     ;Compare X with hex 45
      BNE LOOP     ;Repeat if not zero (i.e. loop 5 times)
```

This program adds 5 numbers into accumulator A. The 5 numbers are held in page 0 of memory, at locations 40 to 44. Register X is used both to index the memory address and as a loop count.

TABLE 4.1 *MC6800 Instruction Set*

Mnemonic	Description
Data move	
LDA A source	Load accumulator A from source (various addressing modes)
LDA B source	Load accumulator B from source
STA A dest	Store accumulator A in destination
STA B dest	Store accumulator B in destination
LDX source	Load register X from source
STX dest	Store register X in destination
TAB	Transfer accumulator A to accumulator B
TBA	Transfer accumulator B to accumulator A
TAP	Transfer accumulator A to status register
TPA	Transfer status register to accumulator A
TXS	Transfer index register X to stack pointer and decrement
TSX	Transfer stack pointer to index register X and increment
LDS data	Load stack pointer immediate
STS dest	Store stack pointer (various addressing modes)
Data modify (A can be replaced by B)	
ABA	Add accumulator B to accumulator A
ADC A source	Add to A with carry
ADD A source	Add to A
AND A source	AND with A
ASL source	Arithmetic shift left (various addressing modes)
ASL A	Arithmetic shift left A
ASR source	Arithmetic shift right (various addressing modes)
ASR A	Arithmetic shift right A
BIT A source	AND A with source (only status bits are affected)
CBA	Compare accumulators
CLC	Clear carry flag
CLI	Clear interrupt mask
CLR source	Clear source
CLV	Clear overflow
CMP A source	Compare source to A
CPX data	Compare immediate with X
COM source	Complement (invert all bits) source
DAA	Decimal adjust A (not available for B)
DEC source	Decrement source
DEC A	Decrement A
DEX	Decrement X
DES	Decrement stack pointer
EOR A source	Exclusive OR A
INC source	Increment source
INC A	Increment A
INX	Increment X
INS	Increment stack pointer
LSR source	Shift right logical source
LSR A	Shift right logical A
NEG source	Negate source
NEG A	Negate A
ORA A source	OR A with source
ROL source	Rotate left through carry source
ROL A	Rotate left through carry A
ROR source	Rotate right through carry source
ROR A	Rotate right through carry A

continued

TABLE 4.1 *Continued*

Mnemonic		Description	
SEC		Set carry	
SEI		Set interrupt mask	
SEV		Set overflow	
SUB	A source	Subtract source from A	
SBA		Subtract B from A	
SBC	A source	Subtract with carry source from A	
TST	source	Test source for zero or minus	
TST	A	Test A for zero or minus	
PSH	A	Push A to stack	
PUL	A	Pull (pop) A from stack	
Branch/jump			
BRA	displacement	Branch always (relative)	
BCC	displacement	Branch if carry clear	
BCS	displacement	Branch if carry set	
BEQ	displacement	Branch if equal to zero	
BGE	displacement	Branch if greater than or equal to zero	
BGT	displacement	Branch if greater than zero	
BHI	displacement	Branch if higher	
BLE	displacement	Branch if less than or equal to zero	
BLS	displacement	Branch if lower or same	
BLT	displacement	Branch if less than zero	
BMI	displacement	Branch if minus	
BNE	displacement	Branch if not equal to zero	
BVC	displacement	Branch if overflow clear	
BVS	displacement	Branch if overflow set	
BPL	displacement	Branch if plus	
BSR	displacement	Branch to subroutine	
JMP	address	Jump to address	Indexed if byte address,
JSR	address	Jump to subroutine	Extended if double-byte address
RTI		Return from interrupt	
RTS		Return from subroutine	
Control/miscellaneous			
SWI		Software interrupt	
WAI		Wait for interrupt	
NOP		No operation	

Program Example 2

```
        LDX #$50       ;Load X with hex 50
REPEAT: LDA A 0,X      ;Load A with contents of location (50 to 5F)
        INX            ;Increment X
        STA A $0F,X    ;Store A in location (60 to 6F)
        CPX #$60       ;Compare X with hex 60
        BNE REPEAT     ;Repeat until 16 bytes are transferred
```

This program moves 16 bytes of data from memory locations 50 through 5F to memory locations 60 through 6F.

4.5 Support Devices

There are two almost identical versions of PIO offered by Motorola to support their MC6800 series CPUs—the MC6820 PIA ("peripheral interface adaptor") and the MC6821 PIA. The pin functions of these devices are the same, and due to their close similarity only the MC6821 is described here.

The circuit connections for the MC6821 are illustrated in Fig. 4.4. There are two 8-bit ports, each with its own separate interrupt line and two control signals. The first of these control lines sets the interrupt flag in the control register of the PIA, whilst the second can be set to be either an output signal or an interrupt input; often these control lines are not used. There are only two address lines connected to the CPU's address bus, and these give four addresses. However, there are six addressable locations on the PIA, which are obtained as follows (assuming that the device is decoded at hex 8000):

Address

8000	Port A Data buffer (if bit2 = 1 on address 8002)
	Port A Data direction register (if bit2 = 0 on address 8002)
8001	Port B Data buffer (if bit2 = 1 on address 8003)
	Port B Data direction register (if bit2 = 0 on address 8003)
8002	Port A Control register
8003	Port B Control register

Effectively therefore there are three addressable locations for each port:

(a) port buffer itself—through which 8 bits are transferred in or out

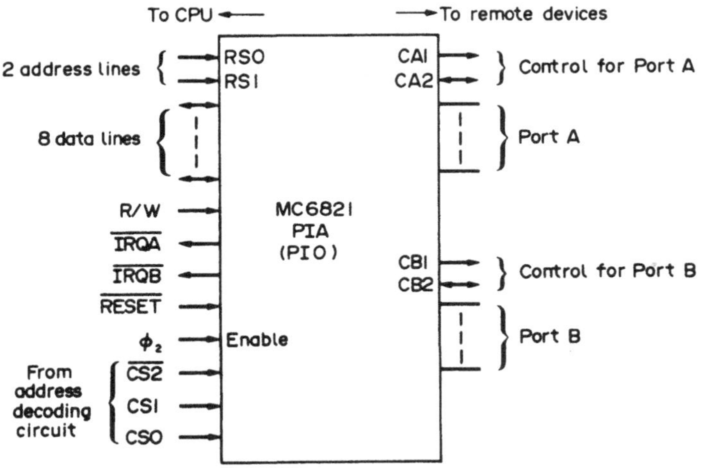

FIG. 4.4. MC6821 PIO circuit connections.

(b) data direction register—allows the programmer to specify the direction of each bit in (a), 0 = output and 1 = input
(c) control register—to select various interrupt options, and (using bit2) to select (a) or (b) above.

If a port is initialised such that the setting of its control line (e.g. CA2) causes an interrupt, and several peripheral devices share the same CPU IRQ interrupt line, then the CPU must poll the interrupting devices to find out which has caused the interrupt. There is no interrupt "daisy-chain" system as with the Zilog Z80.

Motorola also offer a UART to support their CPUs. They name the device an asynchronous communications interface adaptor (ACIA), and it carries the serial number MC6850. It provides both synchronous and asynchronous serial transfer options in the same way as the Intel 8251, but it is easier to program. The circuit connections are illustrated in Fig. 4.5. Although only one address line is connected, four addresses are obtained as follows:

Address

D000	Status register (for data read)
	Control register (for data write)
D001	Tx data (for data read)
	Rx data (for data write)

The status register indicates the state of data items being transferred in or out, and also indicates transmission errors and the states of the modem

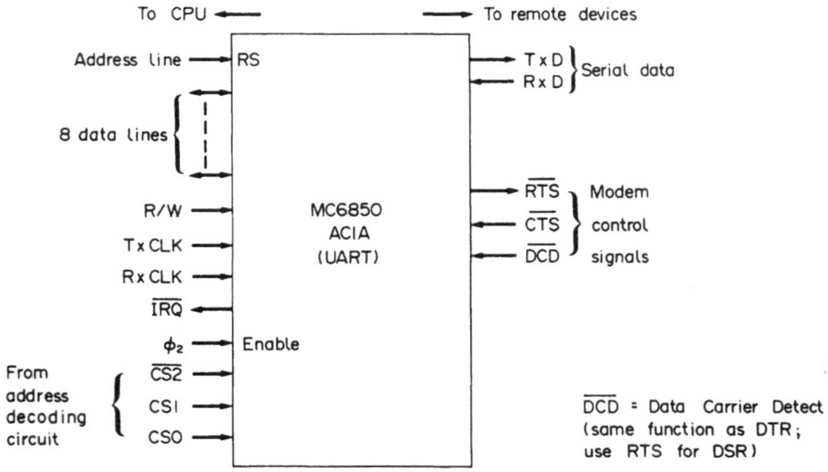

FIG. 4.5. MC6850 UART circuit connections.

control signals. The control register gives the usual RS232-C options, including selection of baud rate. Incoming clock speeds can be divided by 1, 16 and 64 in order to assist in the selection of the required transmit and received baud rates.

Other support devices offered by Motorola are the MC8507 priority interrupt controller (PIC), the XC6852 synchronous serial data adapter (SSDA), the MC6870/MC6871/MC6875 two-phase clocks, the MC6845 CRT controller, and others.

The MC6809 CPU, with its larger instruction set (including some 16-bit operations), is supported by the following two devices which make it suitable for data processing applications:

(a) MC6839 mathematics processor (ROM-based)
(b) MC6829 memory management unit.

4.6 Applications

The MC6809 is used in the Dragon personal computer. Although this particular machine has had limited success, it has considerable computing power for an 8-bit microcomputer configuration.

An interesting application of one of the MC6800 series of CPUs is shown in the simplified schematic of the microprocessor-based VDU (computer terminal) shown in Fig. 4.6. The VDU circuit is based on an MC6808 CPU, which employs the following support devices:

(a) 1K × 8 ROM (MC6830)—holds the program that receives and transmits characters via the serial RS232-C data link, and places received characters in the 8K display RAM
(b) 128 × 8 RAM—display memory to hold ASCII characters received from computer and to be displayed on CRT
(c) 8K × 8 RAM—display memory to hold ASCII characters received from computer and to be displayed on CRT
(d) UART (MC6850)—to transmit and receive characters
(e) keyboard encoder—to detect the operation of a key on the terminal's keyboard
(f) CRT controller (MC6845)—to examine the screen characters stored in (c) and to generate the video signal for application to the CRT.

The central module within the VDU circuit is the 8K RAM display memory. Firstly it is filled by the control program when characters are received from the remote computer through the UART. Secondly it is examined by the CRT controller, which extracts each character-byte in turn for each character positioned along each line on the CRT. As each character is presented to the character generator ROM a different row of dots in the 7 × 9 dot matrix

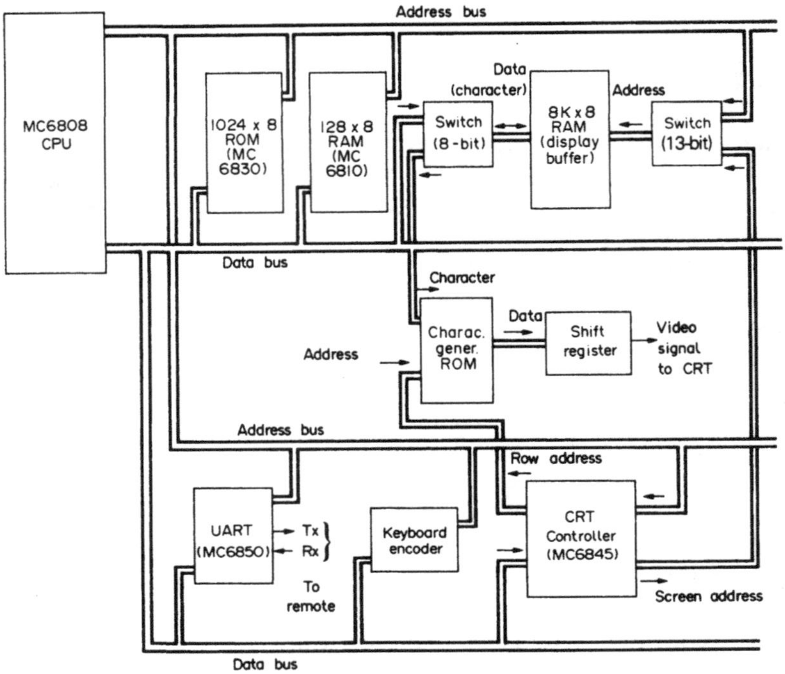

FIG. 4.6. MC6808-based VDU (computer terminal). *Note*: Address decoding and CPU control bus lines not shown, for simplicity.

pattern is addressed in the ROM for each unique character code. The resulting row of dots is converted into a serial bit stream in the shift register. Thus one stream of dots is produced for the top row in a horizontal line of characters, then repeated for the next row for the same line of characters, until all nine rows are formed. The next descending line of characters is then processed in the same manner. When sync pulses are added to this bit stream, a video signal is produced which can be applied to the CRT. This process of video signal generation is continuous. The CPU only sets the changeover switches so that it can address the 8K RAM display area when a new character is received from the remote computer.

Bibliography

1. *Basic Microprocessors and the 6800*. Ron Bishop. Hayden, 1985.
2. *MC6809-MC6809E 8-bit Microprocessor Programming Manual*. Motorola, 1983.

CHAPTER 5

MOS TECHNOLOGY MCS6500 FAMILY (8-BIT)

5.1 Introduction

The MOS Technology MCS6500 range of microprocessors represent enhancements on the Motorola MC6800 range in the same way as the Zilog Z80 is an enhanced form of the Intel 8080 and 8085 devices. There is no machine code compatibility between the MOS Technology and Motorola devices, but CPU architecture is similar and some support devices can be used with both types of microprocessors.

Whilst the other manufacturers of 8-bit microprocessors (Intel, Zilog and Motorola) offer just one or two devices in their range, the MOS Technology MCS6500 CPU family consists of ten devices, as follows:

(a) MCS6502—40-pin, 16 address bus lines
(b) MCS6503—28-pin, 12 address bus lines
(c) MCS6504—28-pin, 13 address bus lines
(d) MCS6505—28-pin, 12 address bus lines
(e) MCS6506—28-pin, 12 address bus lines
(f) MCS6507—28-pin, 13 address bus lines
(g) MCS6512—40-pin, 16 address bus lines
(h) MCS6513—28-pin, 12 address bus lines
(i) MCS6514—28-pin, 13 address bus lines
(j) MCS6515—28-pin, 12 address bus lines.

The MCS6512 is the only device in this range in which the data bus can be floated by the setting of an external signal. All devices possess on-chip clock logic. The ten CPUs share the same instruction set and addressing modes. The advantage of this large family of devices, which differ only in packaging size and memory addressing range, is that a specific device can be chosen to suit a particular application. By far the most popular CPU in the range is the MCS6502, which forms the main processing chip in the following common home and office computers:

(1) Commodore PET
(2) Commodore VIC 20

(3) Apple 2
(4) BBC Model B
(5) Acorn Atom
(6) Oric 1.

Second sources to MOS Technology for the MCS6500 devices are Synertek and Rockwell International.

5.2 MCS6502 Pin Functions

Only the 6502 CPU will be examined in this section due to its overwhelming popularity compared with the other devices in the MCS6500 family. Figure 5.1 shows its pin functions. A single +5 V power supply is required. The input clock signal (fed from a crystal or an RC circuit) is ϕ_0, whilst ϕ_1 and ϕ_2 are output clock signals. Apart from RESET there are two interrupt signals—$\overline{\text{IRQ}}$ (maskable) and $\overline{\text{NMI}}$ (non-maskable). The CPU sets the signal SYNC when the opcode fetch operation is occurring, whilst SO sets the overflow flag. The RDY (Ready) input signal can be utilised with slow

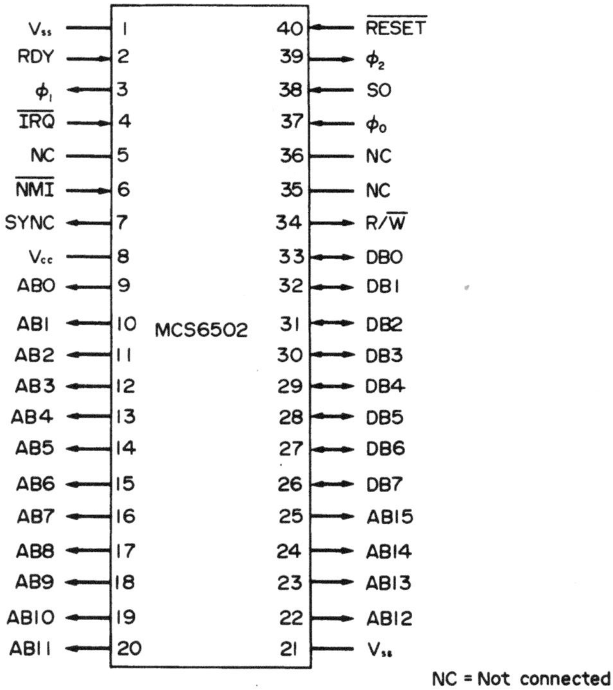

FIG. 5.1. MCS6502 pin functions.

memory systems, and when it is set the CPU inserts a wait machine cycle. $R/\overline{W}$ is the standard read/write signal.

5.3 Processor Architecture

The register organisation for all MCS6500 series CPUs is shown in Fig. 5.2. There are only three registers (A, X and Y) in which data items can be manipulated (cf. seven for the Intel 8080/8085 and Zilog Z80), and so more frequent dumping of data items into memory may be necessary in long program sections compared with other microprocessors. The index registers can be used very flexibly to assist in accessing data bytes, and this is described in the following section (5.4).

The stack pointer is only 8 bits wide, which means that the stack is limited to 256 bytes. However this is unlikely to be a limitation because few applications would require stack expansion beyond this figure. Figure 5.2 shows how a fixed binary 1 is assumed for the ninth bit (most significant) for every use of the stack pointer. This implies that the address range for the stack is hexadecimal 100 to 1FF (denary 256 to 511), as illustrated in Fig. 5.3. The diagram illustrates the memory "paging" system. Page 1 is reserved for the stack, whilst page 0 is conveniently used for data storage, because the instruction set allows 8-bit addressing within this page. For this reason page 0 is sometimes regarded as the "working register" area, because it extends the CPU register space by 256 bytes (with an additional byte within the instruction used to specify the particular location within page 0). Normal 16-bit addressing can be considered to be split into "page" and "location within page" as shown in Fig. 5.4. Whenever an instruction memory access crosses a page boundary an extra cycle delay (one clock pulse) occurs.

Figure 5.3 also shows that the bottom few locations (highest addresses) are reserved for the interrupt vectors. For example, if $\overline{NMI}$ interrupt signal is set then the CPU examines memory locations FFFB and FFFA to find the start address of the interrupt service routine, i.e. places the 16-bit address it finds in these locations into the program counter. For this reason the highest addresses which include these interrupt vectors are ROM-based, whilst pages 0 and 1 are RAM-based in most applications.

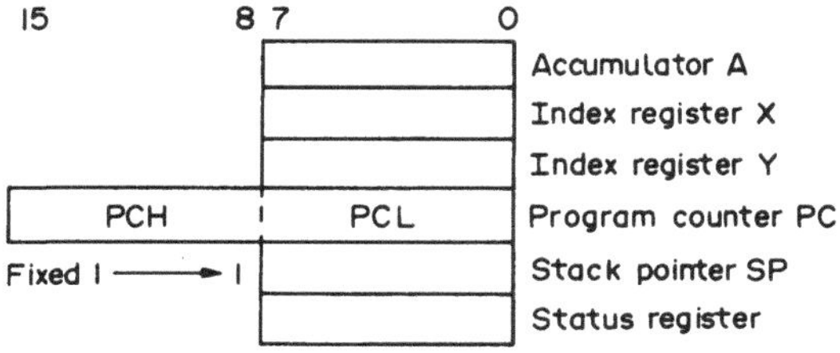

FIG. 5.2. Register set for MCS6500 CPUs.

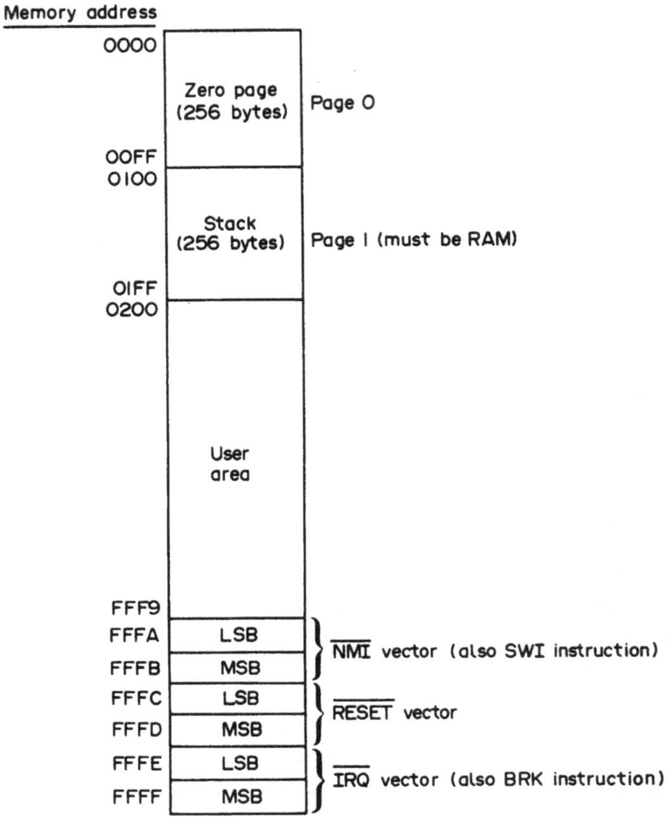

FIG. 5.3. Memory map for MCS6502.

The instruction timing for MCS6500 CPUs is simpler than that for the Intel 8080/8085 and Zilog Z80 devices, which require several clock pulses (typically 4 or 5) for each machine cycle (e.g. memory read operation). A single input clock pulse triggers a single machine cycle.

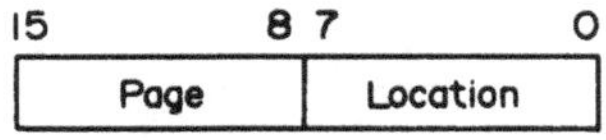

FIG. 5.4. Paged memory address.

5.4 Instruction Set

The MCS6500 instruction set is summarised in Table 5.1, and its has many similarities with the Motorola instruction set. There are no input/output instructions and so input/output must be memory mapped. There are a large number of addressing modes, due to the flexible use of the index registers X and Y, as follows ($ denotes hex):

(a) Register (or accumulator)—MCS literature call this "implied"

`TXA` ;Load A with contents of X

(b) Direct

`LDA $50` ;Load A with contents of memory address 50 in page 0, i.e. address 0050 (2 byte instruction)

(c) Extended Direct

`LDA $4000` ;Load A with contents of memory address 4000 (3 byte instruction)

(d) Immediate

`LDA #$B3` ;Load A with B3 (# indicates immediate addressing mode, and distinguishes from direct mode, as in (b))

(e) Absolute indexed

`LDA $1300,X` ;Load A with contents of memory address computed by adding contents of X to 1300

Register Y can be used in place of X of course.

(f) Base page indexed

`LDA $60,X` ;Load A with contents of memory address computed by adding contents of X to 0060

(g) Indirect pre-indexed

`LDA ($30,X)` ;Load A with contents of memory address held in page 0 address computed by adding contents of X to 30 (ignoring carry)—the 16-bit address is held in (30+X) and (30+X+1)

Note that register Y cannot be used in this mode.

(h) Indirect post-indexed

`LDA ($70),Y` ;Load A with contents of memory address computed by adding the 16-bit memory address held at location 70 (and 71) in page 0 to the contents of Y

Note that register X cannot be used in this mode.

(i) Relative

`BCC +8` ;Branch if carry flag is clear forward 8 bytes

Note that full 16-bit absolute addressing is available with subroutine call and unconditional jump instructions. e.g.

`JMP $2080` ;Jump to memory location 2080

TABLE 5.1 *MCS6500 Instruction Set*

Mnemonic	Description
Data move	
LDA source	Load accumulator from source (various addressing modes)
STA dest	Store accumulator in destination
LDX source	Load X from source
STX dest	Store X in destination
LDY source	Load Y from source
STY dest	Store Y in destination
TAX	Move A to X
TXA	Move X to A
TAY	Move A to Y
TYA	Move Y to A
TSX	Move contents of stack pointer to X
TXS	Move X to stack pointer
PHA	Push A to stack
PLA	Pull (Pop) A from stack
PHP	Push status register
PLP	Pull (Pop) status register
Data modify	
ADC source	Add to A with carry
AND source	AND with A
ASL	Arithmetic shift left A
ASL source	Arithmetic shift left source
BIT source	AND A with source (only status bits are affected)
CLC	Clear carry flag
CLD	Clear decimal mode
CLI	Clear interrupt (enable)
CLV	Clear overflow flag
CMP source	Compare source with A
CPX source	Compare source with X
CPY source	Compare source with Y
DEC source	Decrement source memory location
DEX	Decrement X
DEY	Decrement Y
EOR source	Exclusive OR immediate with A
INC source	Increment contents of source
INX	Increment X
INY	Increment Y
LSR	Logical shift right contents of A
LSR source	Logical shift right contents of source
ORA source	OR contents of source with A
ROL	Rotate left A through carry
ROL source	Rotate left source through carry
SBC data	Subtract immediate from A with borrow
SEC	Set carry flag
SED	Set decimal mode
SEI	Set interrupt (disable)
Branch/jump	
BCC displacement	Branch on carry clear
BCS displacement	Branch on carry set
BEQ displacement	Branch if equal to zero
BMI displacement	Branch on minus

TABLE 5.1 *Continued*

Mnemonic		Description
BNE	displacement	Branch if not equal to zero
BPL	displacement	Branch on plus
BVC	displacement	Branch on overflow clear
BVS	displacement	Branch on overflow set
JMP	address	Jump to address
JSR	address	Jump to subroutine
RTI		Return from interrupt
RTS		Return from subroutine
Control/miscellaneous		
BRK		Break (software interrupt—uses vector at FFFE & FFFF)
NOP		No operation

The following program examples illustrate use of the instruction set (notice the similarity of these program instruction mnemonics with those for the MC6800 in the last chapter):

Program Example 1

```
     LDY #6      ;Load Y with count of 6
     LDA #0      ;Clear A
LOOP:STA $50,Y   ;Store A in data table (51 to 56, page 0)
     DEY         ;Decrement Y
     BNE LOOP    ;Repeat if not zero (i.e. loop 6 times)
```

This program clears 6 memory locations held in page 0. Notice that register Y is used both as a loop counter and to index the memory address of the data destination in memory.

Program Example 2

```
      LDA #1       ;Load A with 1
      STA $6000    ;Output to LED (address hex 6000)
      LDX #$FFFF   ;Set delay count in X
DELAY:DEX          ;Decrement X
      BNE DELAY    ;Repeat until X is zero
      LDA #0       ;Load A with 0
      STA $6000    ;Output to LED (address hex 6000)
      LDX #$FFFF   ;Second
PAUSE:DEX          ;        delay
      BNE PAUSE    ;             loop
      JMP START    ;Repeat program continuously
```

This program continuously flashes a LED, which is "memory-mapped" at address 6000. A time delay of typically 0.5 s (depending on the CPU clock speed) is achieved for both LED "on" and "off" times using a 16-bit delay count in X register.

5.5 Support Devices

MOS Technology offer two versions of a PIO—the MCS6520 peripheral interface adaptor (PIA) and the MCS6522. The latter device is an enhanced form of the former, and it offers more handshaking logic on Port A (both devices support two ports) and two counter/timer circuits. For these reasons the MCS6522 only is described here.

Figure 5.5 shows the circuit connections for the MCS6522. The second of the two control signals associated with each port can be set to be input or output in direction. This facilitates the connection of a parallel printer via the Centronics interface (8 data bits plus 2 handshaking signals—Strobe to printer, Acknowledge or Busy from printer), which is one of the most common applications for PIO ports. The four address lines give 16 addresses on the device, as detailed in Fig. 5.6. Before the two ports are applied for data transfer their directions must be set by loading addresses 2 and 3 with direction identification bytes. The peripheral control register is used to define the mode of operation of the control signals associated with each port, e.g. polarity (rising edge or falling edge) of input control line. The auxiliary control register specifies if the port inputs are to be latched, and it also controls the T1 and T2 counters. These counters are 16 bits wide, and can be

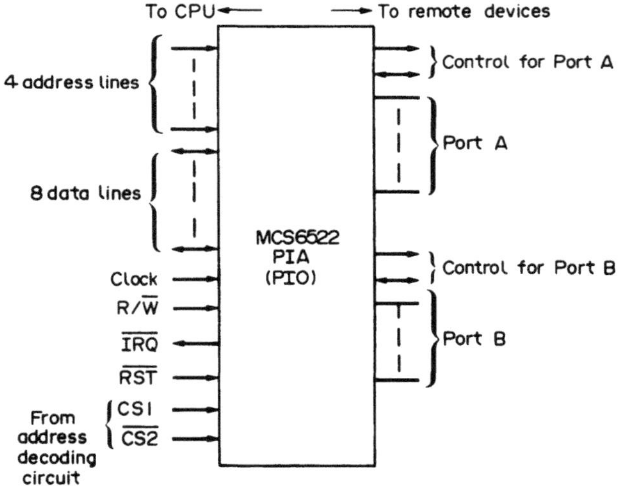

FIG. 5.5. MCS6522 PIO circuit connections.

Address	Identity
0	Port B
1	Port A
2	Direction for Port B
3	Direction for Port A
4	T1 low-order counter
5	T1 high-order counter
6	T1 low-order latches
7	T2 high-order latches
8	T2 low-order counter
9	T2 high-order counter
A	Shift register
B	Auxiliary control register
C	Peripheral control register
D	Interrupt flag register
E	Interrupt enable register
F	Port A (no handshaking)

FIG. 5.6. MCS6522 addressable registers.

operated in "one-shot mode" (decrement only once) or "free-running mode" (counter is automatically reloaded and repeats countdown).

Another popular support device for MCS6500 CPUs is the UART offered by Synertek and named the SY6551 asynchronous communications interface adaptor (ACIA). The SY6551 is an enhancement of the Motorola MC6850, and can be applied with both MC6800 and MCS6500 series microprocessors. Figure 5.7 shows the circuit connections. The standard RS232-C serial data channels are TxD and RxD, whilst the usual range of modem control signals

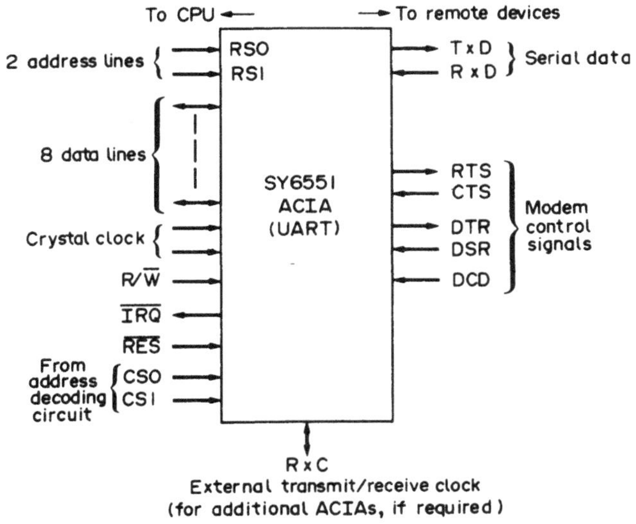

FIG. 5.7. SY6551 UART circuit connections.

TABLE 5.2 *SY6551 Addresses*

RS1	RS0	R/$\overline{W}$	Identity
0	0	1	RxD receive data
0	0	0	TxD transmit data
0	1	1	Status register (read)
0	1	0	Reset
1	0	1/0	Command register (read or write)
1	1	1/0	Control register (read or write)

are available. An extremely wide range of baud rates is achievable, viz. 50 baud to 19200 baud, assuming a crystal frequency of 1.8432 MHz. There are six addressable locations within the device, as shown in Table 5.2.

The control register is used to specify baud rate, number of data bits and number of stop bits. The command register specifies parity and interrupt options. The status register indicates errors (parity, framing, overrun) and receive and transmit buffers state (full or empty).

5.6 Applications

Figure 5.8 shows a simplified schematic of the BBC Model B microcomputer. The MCS6502 CPU feeds its buses to 32K ROM (supporting the BASIC interpreter and the operating system, or organiser program) and

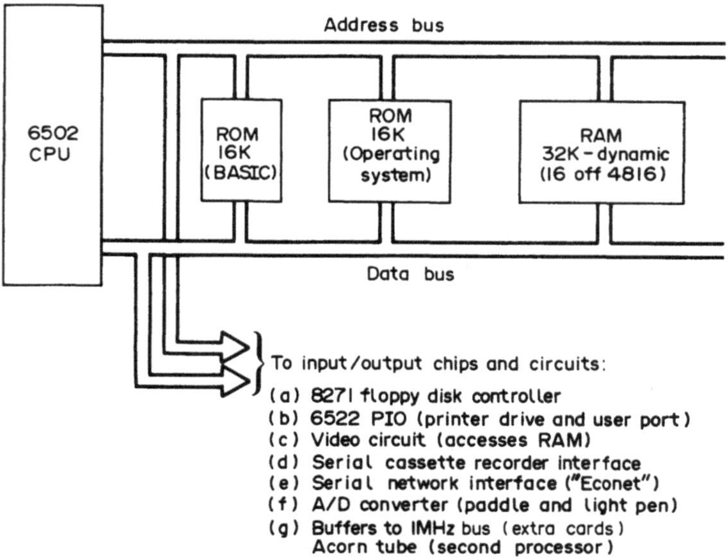

FIG. 5.8. Simplified schematic of BBC Model B microcomputer.

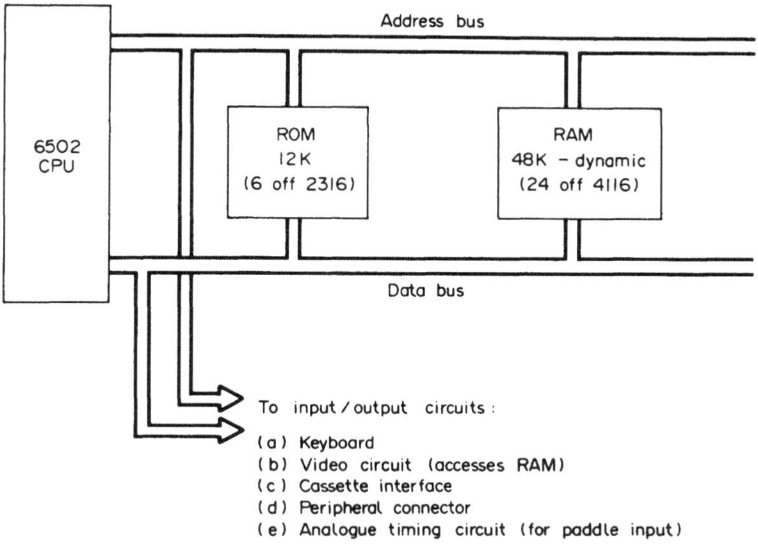

FIG. 5.9. Simplified schematic of Apple II microcomputer.

32K RAM (video graphics and user area). The buses also drive a variety of input/output devices and circuits including floppy disc, cassette recorder, printer, games input (e.g. paddle) and facilities to connect to additional circuit boards and other computers.

A simplified circuit arrangement of the Apple II microcomputer (main board) is illustrated in Fig. 5.9. The operating system is not ROM-based and must be loaded ("bootstrapped") from floppy disc into RAM on start-up. A second board handles the drive to the floppy disc unit.

5.7 Multi-function Support Devices

There are three chips which represent an attempt to perform the role of a single-chip microcomputer based on the 6500 architecture, but they are not full single-chip microcomputers in the true sense. They do not possess CPU architecture and cannot therefore operate as single-chip systems. Rather they are multi-function input/output devices, which can be used to support an MCS6500 series CPU, as follows:

(a) MCS6530, which provides:
  (1) 1K bytes ROM
  (2) 64 bytes RAM
  (3) 2 input/output ports
  (4) programmable interval timer
  (5) interrupt logic.

(b) MCS6532, which is almost identical to the MCS6530 except that there is no ROM, but RAM is 128 bytes.
(c) Rockwell R6531, which represents a significant enhancement over the two previous devices, and provides:
   (1) 2K bytes ROM
   (2) 128 bytes RAM
   (3) 4 input/output ports (Port C output, Port D input only)
   (4) programmable interval timer
   (5) interrupt logic
   (6) primitive serial input/output (Port B, pin 3).

Bibliography

1. *Osborne 4 & 8-bit Microprocessor Handbook.* Adam Osborne and Gerry Kane. Osborne/Mcgraw-Hill, 1981.
2. *6502 Applications Book.* Rodney Zaks. Sybex, 1979.
3. *6502 Assembly Language Programming.* Lance A. Leventhal. Osborne/McGraw-Hill, 1979.

CHAPTER 6

INTEL 8086/8088 FAMILY (16-BIT)

6.1 Introduction

Intel were the first firm to offer 16-bit microprocessors. The 16-bit 8086 represents a major upgrade on the 8-bit 8080 and 8085 devices, whilst the 8088 is virtually identical to the 8086 except that its external data bus is only 8 bits wide.

Intel carry the main share of the 16-bit microprocessor market—figures of between 70% and 80% are quoted. The 8086 and 8088 dominate in the 16-bit office computer market, as follows:

Machine	Microprocessor
IBM personal computer (early versions)	8088
ACT Sirius	8088
DEC Rainbow	8088
Apricot	8086

Notice that later versions of the IBM PC (and imitations, or "clones") use enhanced versions of the 8088/8086. These versions are the 80186 and 80286, which are described in sections 6.8 and 6.9, and the 80386, which is described in chapter 9.

The principal advantages that the Intel 16-bit microprocessors, just like 16-bit devices offered by other manufacturers (Motorola, Zilog and National Semiconductor), offer compared with 8-bit devices are:

(a) faster instruction execution times
(b) wide instruction set (e.g. multiplication and divide instructions)
(c) larger memory addressing range—typically 1M byte (or more) compared with 64K byte
(d) larger integer number range (0–64K in place of 0–255)
(e) more addressing modes—to make programs simpler and more efficient
(f) the use of co-processors to assist the CPU in executing programs faster.

6.2 8086/8088 Pin Functions

Figures 6.1 and 6.2 show the pin functions of the 8086 and 8088 DIL packages. They are both 40-pin devices with a single +5 V power/ground supply. Notice the following features:

(a) The 16-bit address and data buses are multiplexed (share the same pins) on the 8086. Only the bottom 8 bits of the address bus are multiplexed with the 8088 because the 8088 has only an 8-bit external data bus. The signal ALE (Address Latch Enable) identifies whether a memory address or data is being transferred.

(b) There are 20 address bus lines in each case—this gives 1M byte memory addressing range.

(c) The 40-pin limitation in DIL size caused Intel to multiplex some signals in the control bus (in addition to multiplexing the address and data buses). The diagrams show that 8 control bus lines (pins 24 to 31) have dual functions, as selected by the setting of the MN/$\overline{MX}$ signal. This signal selects "Minimum" mode if connected to +5 V, and the 8 identities ($\overline{INTA}$ to HOLD), which are closely related to functions for 8-bit microprocessors, are selected. If MN/$\overline{MX}$ is connected to Ground, then "Maximum" mode is selected, which causes an expansion of the 8 control signals to facilitate connection to large and complicated circuit configurations. This expansion is created because 3 signals ($\overline{S0}$, $\overline{S1}$ and $\overline{S2}$) are connected to a 3 to 8 decoder (the Intel 8288 bus controller) to provide 8 discrete control bus signals (DEN, DT/R and IO/M are a subset of these 8), and 2 signals (QS1 and QS0) are connected to a 2 to 4 decoder.

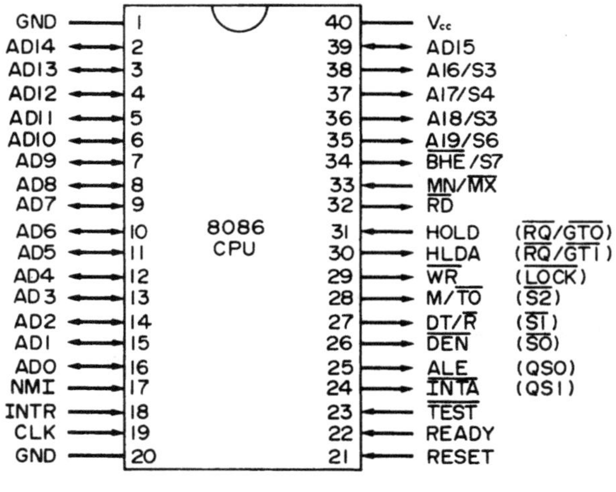

FIG. 6.1. 8086 pin functions.

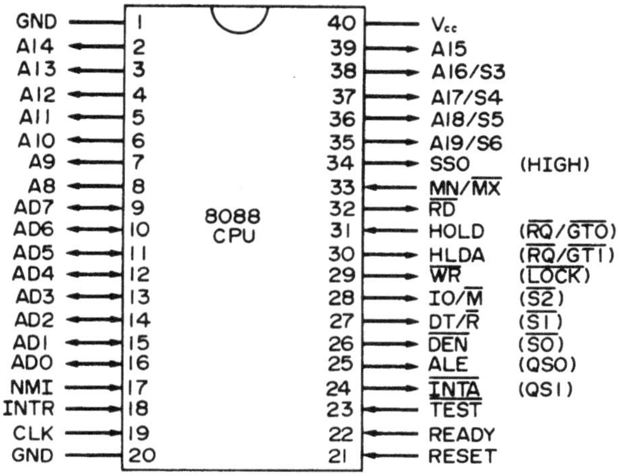

FIG. 6.2. 8088 pin functions.

The standard range of 8-bit input/output chips and memory chips are connected to these 16-bit microprocessors in full circuit configurations (examples are given at the end of this chapter). The multiplexed address and data bus PIOs (the Intel 8155 and 8255) can be connected directly, whilst the 8251 UART can be connected by linking the data pins to the multiplexed ADO to AD7 and its C/D pin to the demultiplexed AO. However, other devices, and in particular memory chips, require that the address and data buses be demultiplexed. This can be achieved using Intel 8282 8-bit latch buffers or Intel 8286 8-bit bi-directional bus transceivers—alternatively the standard TTL SN74245 transceivers can be applied.

One other essential support chip requires mention here. The Intel 8284 clock generator provides the CPU Clock (the input crystal frequency is divided by 3) and Reset signals. A typical Clock rate is 5 MHz.

Variations of the basic 8086 and 8088 have been introduced by Intel in 1982/3 as follows:

(a) iAPX 186 and iAPX 188 (alternatively named the 80186 and 80188)—offer additional instructions and on-chip clock generator, interrupt controller and DMA controller—described in 6.8;
(b) iAPX 286 (alternatively named the 80286)—this does not support the on-chip functions, such as clock generator, of the iAPX 186, but includes yet more instructions, plus the ability to support "virtual memory" (addresses backing store as if it is an extension of main memory)—used in the IBM PC "AT" version and described in 6.9.

6.3 Processor Architecture

The internal operation of the 8086 and 8088 microprocessors is illustrated in Fig. 6.3. The architecture is split into two separate processing units—the execution unit (EU) and the bus interface unit (BIU). The former comprises the main components found in any 8-bit microprocessor—ALU, control unit, instruction register, flags (or status register) and general registers (note that there are 8 16-bit work registers). The bus interface unit processes signal connections to/from the multiplexed address/data bus and control bus, and performs the following functions:

(a) organises the multiplexing timing
(b) automatically fills an instruction queue with the following instruction/instructions
(c) adds the contents of the four segment registers to memory addresses before they are placed in the address bus, such that memory is accessed in 64K byte "segments".

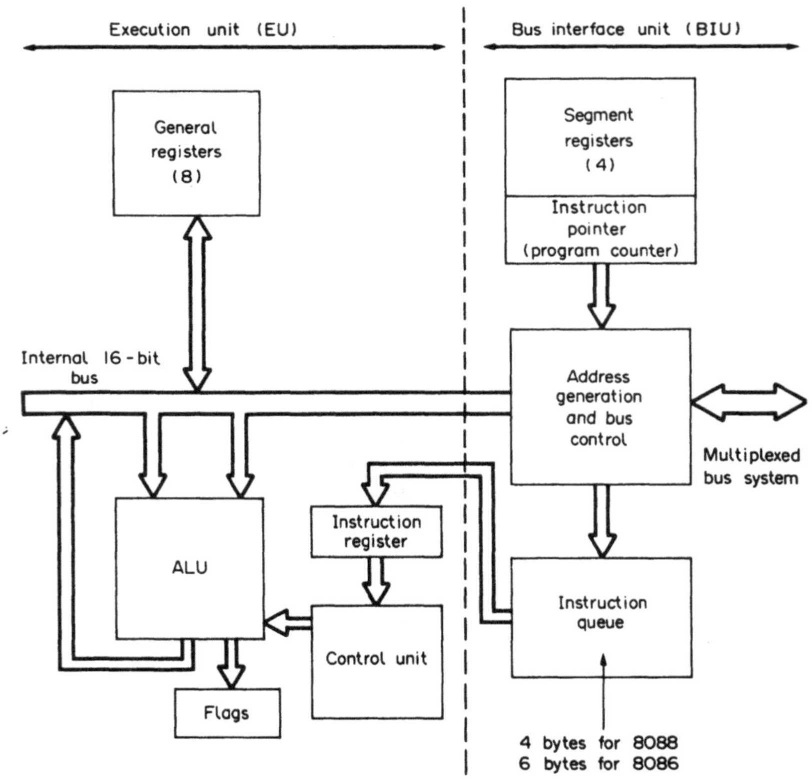

FIG. 6.3. Internal architecture of 8086 and 8088.

The last two features are worthy of more detailed descriptions. The instruction queue is a feature that helps to increase the speed of operation of the device. Memory transfers represent a major time delay in the execution of a sequence of program instructions. Therefore the BIU continually attempts to keep the instruction queue full by fetching succeeding instruction bytes from memory when the EU is executing an instruction but does not require

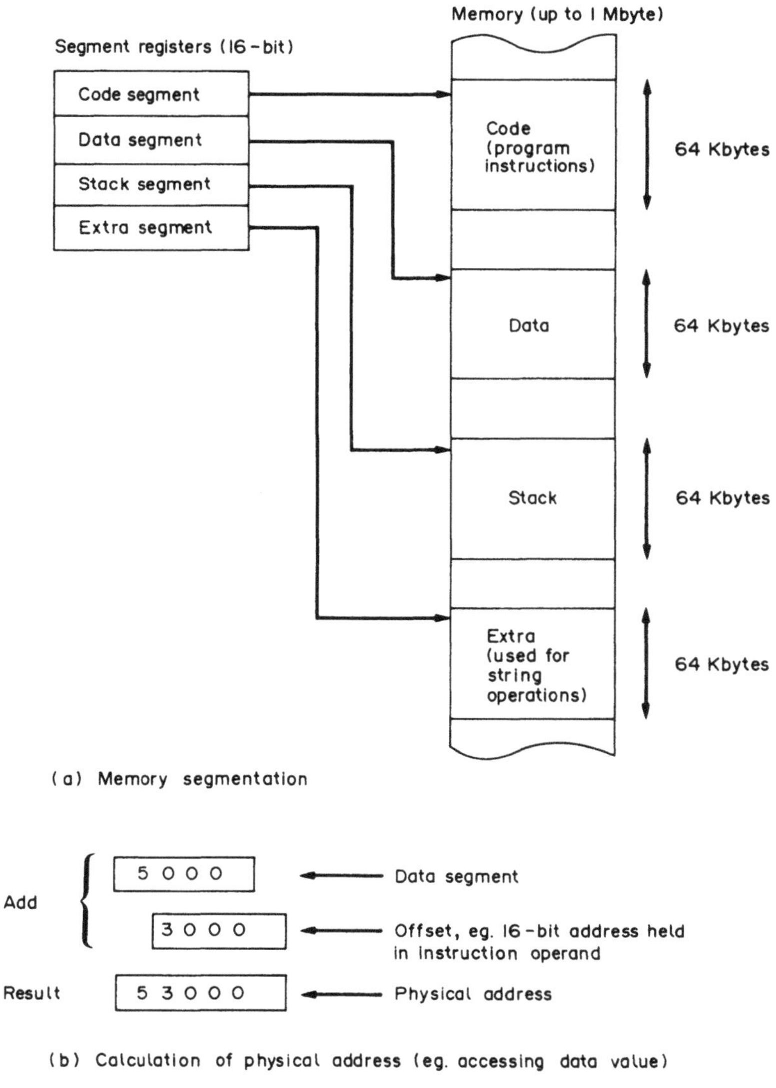

FIG. 6.4. Generation of 20-bit addresses using segment registers.

use of the buses. In this way when the EU completes an instruction it does not have to perform a memory read to access the next instruction opcode. Instructions are variable in length (1, 2, 3, 4 or more bytes), and the instruction queue is 6 bytes long in the 8086 and 4 bytes long in the 8088. If an instruction (e.g. jump or call) that transfers control to another location in the program is obeyed, the BIU clears the queue and passes the instruction from the new address directly to the EU. It then refills the instruction queue. This queue feature is sometimes called an instruction "pipeline", and it ensures more efficient use of the buses and faster program execution times.

The other facility within the BIU that warrants a more detailed explanation is the use of the segment registers. Figure 6.4 shows the identities of the four segment registers, and how they are applied. The 1M byte address range of memory is divided into 64K byte "segments". Different programs can occupy different segments. Similarly different data lists can be placed in different 64K segments, and the stack can reside in its own 64K segment. The contents of a 16-bit segment register are effectively shifted left 4 places and then added to the effective memory address (as used by any 8-bit microprocessor) in the BIU to produce a 20-bit physical memory address that is placed on the address bus. The choice of which segment register is used is implied in the instruction, so that program instructions are fetched from one segment whilst the data values used in that program are accessed from another segment. Segments can overlap, of course, and in a system in which the total amount of memory is less than 64K bytes it is possible to set all segment registers to the same value, e.g. zero, and have totally overlapping segments. Segmentation is extremely useful for large multi-tasking systems to provide isolation between program modules and data lists. Additionally programs are relocatable.

6.4 Instruction Set

The instruction set is based upon manipultion of data values held in the general registers, and Fig. 6.5 lists these eight 16-bit registers. The first four

Data registers	AX	AH	AL	Accumulator
	BX	BH	BL	Base
	CX	CH	CL	Count
	DX	DH	DL	Data
Index registers		SP		Stack pointer
		BP		Base pointer
		SI		Source index
		DI		Destination index

FIG. 6.5. 8086/8088 general registers.

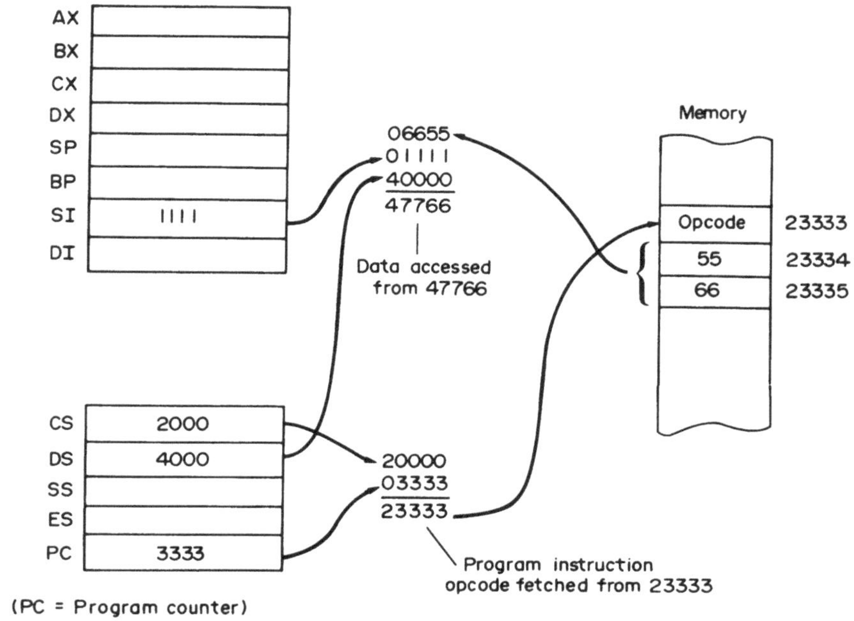

FIG. 6.6. Memory address calculations (source is direct, indexed addressing).

registers are the principal data manipulation registers, and they can also be accessed in halves (bytes). All four registers act as accumulators, whilst BX can serve as a base register when computing data memory addresses, CX can be used as a counter in multi-iteration instructions and DX can be used to transfer data to input/output addresses (ports).

The stack pointer SP operates in the normal way, but the stack segment register is added to it to access the required memory segment, as described in the previous section. The base pointer BP can be used in some instructions to provide a 16-bit base to a computed memory address. The source and destination index registers can be used to add a displacement to a memory address that is applied in either the source or destination within an instruction. Figure 6.6 illustrates in detail the method which is applied to access a program instruction, and to access the 16-bit data item that is used within that instruction, for "direct, indexed addressing". The instruction set is listed in Table 6.1, and the addressing modes available are:

(a) Immediate
```
MOV AX,29          ;Load AX with 29
```
(b) Register
```
MOV AX,BX          ;Load AX with contents of BX
```

(c) Direct

```
MOV AX,1040H            ;Load AX with contents of memory
                         address at offset 1040
```

(d) Indirect, with base register

```
MOV AX,(BX)             ;Load AX with contents of memory
                         address held in BX
```

(e) Indirect, with index register

```
MOV AX,(SI)             ;Load AX with contents of memory
                         address held in SI
```

(f) Indirect, with base register plus index register

```
MOV AX,(BX)(SI)         ;Load AX with contents of memory
        ↑   ↑            address computed by summing BX
    or BP  or DI         and SI
```

(g) Indirect, with base register *or* index register, plus offset

```
MOV AX,4000H(BP)        ;Load AX with contents of memory
                         address computed by summing BP
                         and 4000
```

(h) Indirect, with base register plus index register plus offset

```
MOV AX,2100H(BX)(SI)    ;Load AX with contents of memory
                         address computed by summing BX,
                         SI and 2100
```

(i) Relative

```
JMP -23                 ;Jump (unconditional) to PC-23, i.e.
                         back 23 bytes
```

TABLE 6.1 *8086/8088 Instruction Set (main instructions)*

Mnemonic		Description
Data move		
MOV	dest,source	Move data value (byte or 16-bit word) from source to destination (both can be a register or memory location, or immediate value for source)
LDS	reg,source	Load pointer using data segment
LEA	reg,source	Load effective address into register
LES	reg,source	Load pointer using extra segment
XCHG	dest,source	Exchange contents of register, or of register with memory
XCHG	AX(or AL), reg	Exchange register and accumulator
IN	AL,port	Input from port address to AL (or AX)
IN	AL	Input from port address held in DX to AL (or AX)
OUT	port,AL	Output to port address from AL (or AX)
OUT	AL	Output to port address held in DX from AL (or AX)
Usual register PUSH and POP instructions		
Data modify		
ADD	dest,source	Add source to destination
ADC	dest,source	Add with carry
SUB	dest,source	Subtract

TABLE 6.1 *Continued*

Mnemonic		Description
SBB	dest,source	Subtract with borrow
AND	dest,source	AND
TEST	dest,source	AND with no result obtained
OR	dest,source	OR
XOR	dest,source	Exclusive-OR
INC	dest	Increment (register or memory)
DEC	dest	Decrement
AAA		ASCII adjust for add
DAA		Decimal adjust for add
NEG	dest	Negate (two's complement)
AAS		ASCII adjust for subtract
DAS		Decimal adjust for subtract
MUL	source	Multiply accumulator (by register or memory)
DIV	source	Divide accumulator (by register or memory)
IMUL	source	Signed multiply
IDIV	source	Signed divide
AAM	dest	ASCII adjust for multiply
AAD	dest	ASCII adjust for divide
CBW		Convert byte to word
CWD		Convert word to double-word (in AX or DX)
NOT	dest	Invert (one's complement)
SHL(SAL)	dest	Shift left by count in CL register
SHR	dest	Shift logical right by count in CL register
SAR	dest	Shift arithmetic right by count in CL register
ROL	dest	Rotate left by count in CL register
ROR	dest	Rotate right by count in CL register
RCL	dest	Rotate left through carry
RCR	dest	Rotate right through carry
Branch/jump		
Jcc	offset	Jump conditional relative to PC (with variety of conditions, e.g. JLE −63)
JMP	address	Jump to 16-bit address within segment or intersegment
JMP	offset	Jump relative to PC (−128 to +127 bytes)
JMP	source	Jump using indirect addressing within segment or intersegment
CALL	address	Call within segment or intersegment
CALL	source	Call using indirect addressing within segment or intersegment
RET		Return
RET	data	Return and add immediate data to SP
LOOP	offset	Loop CX times
LOOPZ	offset	Loop whilst CX is non-zero and zero flag is not set
LOOPNZ	offset	Loop whilst CX is non-zero and zero flag is set
Control/miscellaneous		
String operations instructions, e.g.		
REP		Repeat string operation decrementing CX until CX is zero
MOVSB		Move source string to destination string (source is addressed by SI, destination by DI)
Various control instructions, e.g.		
INT	data	Software interrupt (data represents type)
CLC		Clear carry flag
WAIT		Wait until TEST pin is set

Additional normal absolute address jump/call instructions are available, e.g.

```
    JMP 6500H                ;Jump (unconditional) to 6500
```

Naturally AX in these examples could be replaced with any of the other seven general registers, and the source and destination could be interchanged.

Whilst Table 6.1 lists the main instructions that are available, the reader is referred to the bibliography at the end of the chapter for a comprehensive list of the instruction set for the 8086 and 8088. The following short program examples illustrate simple input/output operations and the application of a multiply instruction:

Program example 1

```
  MOV AX,1000H   ;Set data segment
  MOV DS,AX      ;                    to 1000
  IN  AL,2       ;Input from port address 2
  MOV 0000H,AX   ;Store in data segment (offset 0000)—memory
                  address 10000
  OUT 3,AL       ;Output to port address 3
  HLT            ;Halt
```

The program reads in a byte from an input port into register AL, stores that value in memory and then sends that byte to an output port.

An example of a program written for an 8086/8088 assembler is:

Program example 2

```
;Program multiplies two 16-bit values and stores
;answer in 32-bit form in memory
DATA SEGMENT
  NUM1 DW 1234H
  NUM2 DW 4321H
  ANS1 DW ?
  ANS2 DW ?
DATA ENDS
CODE SEGMENT
  ORG 0000H
  ASSUME CS:CODE,DS:DATA   ;Pseudo, to set up segments
  MOV AX,DATA              ;Set up
  MOV DS,AX                ;       data segment
  MOV AX,NUM1              ;Fetch one number
  MUL NUM2                 ;Multiply by other number
  MOV ANS1,AX              ;Store least significant word of answer
```

```
  MOV ANS2,DX          ;Store most significant word of answer
  HLT                  ;Halt
CODE ENDS
END
```

6.5 Memory Connections

Standard 8-bit memory devices, e.g. 2316 ROM, 2716 EPROM and 6116 static RAM, are connected to the 8088 as if it is a normal 8-bit microprocessor. The only requirement is that the address and data buses must be de-multiplexed, as for the 8085. However, the 8086 possesses a full 16-bit external data bus, and although it is a 16-bit word device it addresses memory in byte form. Intel could have designed the 8086 to use a 16-bit opcode and to lose the ability to access byte data values. This would have restricted instruction lengths to 2, 4 or 6 bytes. However, the 8086 maximises memory usage by using byte opcodes and byte memory addressing. The memory connection arrangement is shown in Fig. 6.7. The bottom address

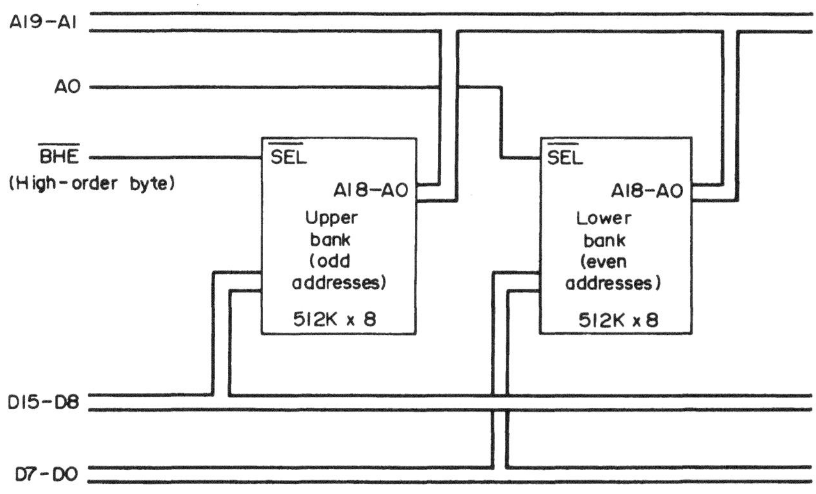

(a) Circuit arrangement (assuming de-multiplexed address and data buses)

$\overline{BHE}$	A0	Byte accessed
0	0	Both bytes – word transfer
0	1	Upper byte (odd address)
1	0	Lower byte (even address)
1	1	Neither

(b) Truth table for $\overline{BHE}$ and A0

FIG. 6.7. 8086 memory connection.

line A0 is not taken to memory devices. Instead it is used as a bank select for lower-order bytes, whilst $\overline{BHE}$ (set by CPU) selects higher-order bytes. In this way both odd and even address memory bytes can be accessed, whilst words (two bytes) are accessed on even address boundaries (top line of truth table) and odd address boundaries.

6.6 Supporting Chips—the 8087 and 8089

Two extremely powerful 40-pin supporting devices are offered by Intel to aid the 8086 and 8088 microprocessors in number processing applications and large circuit configurations. These devices are the 8087 numeric data processor, which offers hardware floating point arithmetic operations, and the 8089 input/output processor, which organises DMA transfers and optimises other input/output operations. Both chips act as co-processors to the CPU, and are connected to the multiplexed address/data bus as well as to selected control bus lines. The CPU must be set to "maximum" mode when these devices are included.

The 8087 monitors the instruction sequence that is being fetched and obeyed by the CPU. If a reserved opcode (11011XYZ) is detected by the 8087, the CPU ignores the instruction and the 8087 executes it in place of the CPU. The 8087 may need to perform memory transfers in order to execute the instruction, and handshaking with the CPU is performed to enable it to "borrow" the CPU buses. The programmer may need to insert CPU wait instructions to ensure that the 8087 has completed its numeric computation task before the result is utilised in the program (there is no hardware indication of task completed). Numeric functions that can be performed are:

(a) high-precision fixed-point arithmetic
(b) floating-point arithmetic
(c) complex mathematical functions, e.g. square root, tangent, exponentiation.

The 8089 does not monitor instructions to determine if it should replace the CPU for certain instructions, like the 8087. Instead the 8089 input/output processor is triggered when the CPU sets a signal pin on the 8089, and a "mailbox" arrangement is applied to enable the CPU to specify the input/output task required. The mailbox is an area of memory that is shared by the CPU and the 8089, and the CPU (by means of a section of program) must load this memory block with both details of the required input/output transfer and the sequence of 8089 instructions. The 8089 executes this instruction block and notifies the CPU, by interrupt, when it has finished the task. The 8089 possesses nearly 50 instructions.

The 8089 input/output processor possesses two "channels", i.e. it can perform the transfer of two separate streams of data. Normally these channels are used for DMA transfers, e.g. to floppy disc and hard disc, but they can be used for a wide variety and flexible range of tasks, e.g. graphics CRT display refresh, data manipulation during block transfers, spooling messages to printers and bus width matching (byte memory for 8088, and 16-bit memory for 8086 with odd-even banks).

An example of the use of both the 8087 and 8089 is given in the following section.

6.7 Example Circuit Configurations

Figure 6.8 shows a small application system based on an 8088 CPU. Four supporting chips offer:

(a) 8284—clock generator

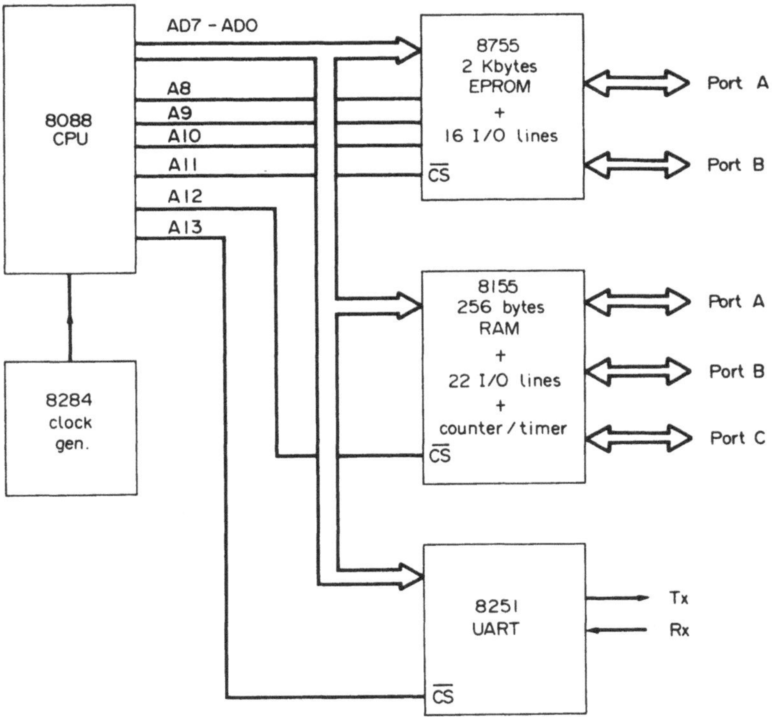

Not shown:

1. Control bus signals, e.g. $\overline{RD}$, $\overline{WR}$, ALE, to 8755 and 8155.
2. Control bus signals, e.g. $\overline{RD}$, $\overline{WR}$ and clock signals, to 8251. Also AD0 de-multiplexed by ALE and connected to C/$\overline{D}$ on 8251.

FIG. 6.8. Small 8088 system.

(b) 8755—PIO with:
2K bytes EPROM
16 input/output lines

(c) 8155—PIO with:
256 bytes static RAM
22 input/output lines
counter/timer

(d) 8251—UART with serial input and output, e.g. to VDU.

The 8755 and 8155 are both multiplexed address/data bus devices and connect directly to the 8085/8086/8088 CPUs. The 8251 is not a multiplexed device, but it can be connected directly to AD7 to AD0 on the 8088, with its single address pin (C/$\overline{D}$) fed by a de-multiplexed A0 address line.

A more complicated circuit arrangement, based on an 8086 CPU, is shown in Fig. 6.9. This simplified representation shows the principal components in

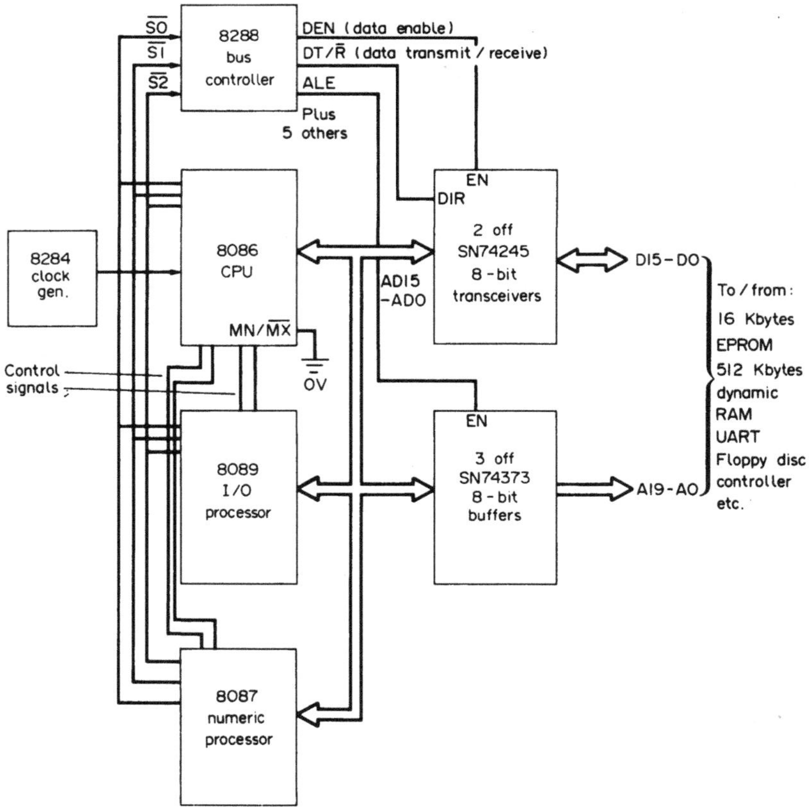

FIG. 6.9. Large 8086 system (Apricot computer).

the Apricot computer. The 8086 is supported by an 8089 input/output processor and an 8087 numeric processor. The CPU is applied in maximum mode, i.e. the control bus is expanded (3 to 8) using the 8288 bus controller. The address/data bus is de-multiplexed using standard TTL chips, and connects to a large range of memory and input/output devices.

6.8 The 80186 CPU

Intel named their updates on the standard 8088 and 8086 the iAPX188 (or 80188) and the iAPX186 (or 80186) when they were launched in 1982. These devices offered some extra instructions and several on-chip facilities that reduced the need for supporting ICs. In this way a microcomputer circuit requiring perhaps ten ICs can be supplied by just one of these devices.

The more popular of these devices is the 80186, which is illustrated in Fig. 6.10. The 8086 architecture is retained as the execution unit (EU) and bus interface unit (BIU)—refer to Fig. 6.3.

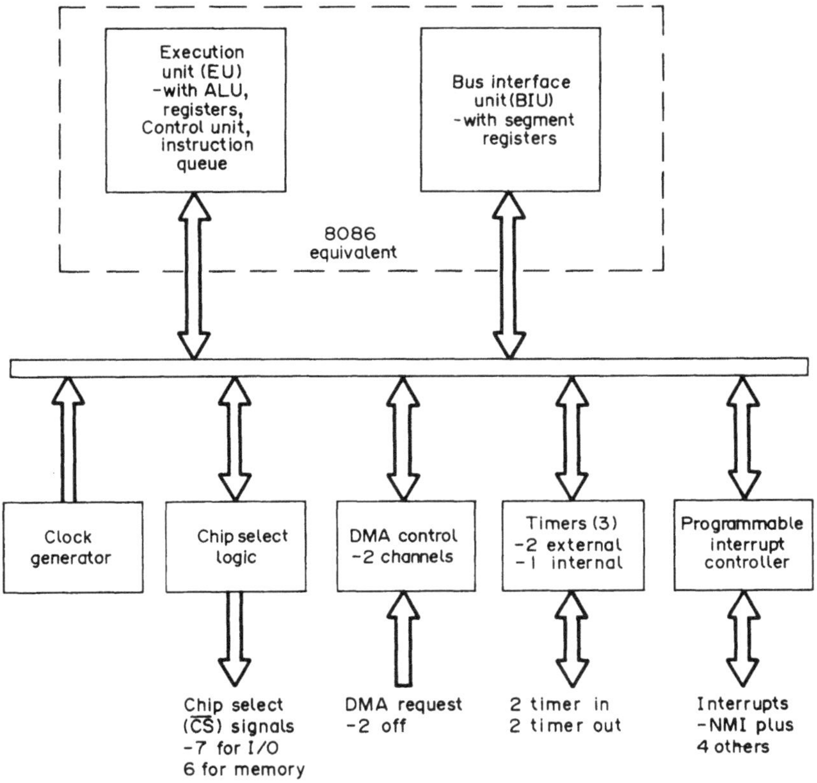

FIG. 6.10. 80186 architecture.

Additional modules are:

(a) clock generator, with programmable wait-state generation logic
(b) chip select logic, avoiding the need for external address decoding circuitry eliminating decoder propagation delays
(c) DMA logic for two channels, with interrupt facilities
(d) programmable timers/event counters, two external and one internal (e.g. for real-time clock)
(e) interrupt controller, which is programmable.

The chip packaging technique is the first one encountered in this book to move away from the standard DIL (dual-in-line) method. The 80186 is packaged in a 68-pin JEDEC type A "chip carrier", and this package arranges its pins on all four sides, allowing a smaller package. The device is targeted at small stand-alone applications employing a low chip-count, but it has not achieved as widespread an application as the device described in the following section.

6.9 The 80286 CPU

Intel introduced the iAPX286 (better known as the 80286) in 1983. Although this device retains compatibility with the 8086 in the same way as the 80186, it offers a radically different range of additional features. It does not offer the on-chip functions of clock generator, chip select logic, DMA control, timers and interrupt controller; instead it provides memory management and virtual memory facilities. Whilst IBM guaranteed widespread application for the 8088 in their first entry into the microcomputer market with the IBM PC (personal computer), they additionally consolidated the popularity of the 80286 by choosing it as the CPU for the updated IBM PC "AT" version.

The concept of virtual memory is that programs and data files are addressed as if they are held in memory, but in fact they could be held on backing store (disc). Thus backing store is treated as if it is an extension of main memory. In a multi-programming system therefore, the master program ("operating system") can switch from one program task to another, perhaps using "time-slicing" (programs are run in short bursts to avoid a lengthy task dominating machine time), without the need to implement disc transfers—transfers are performed automatically by the CPU hardware.

The term "memory management" refers to the process of address translation from the addresses used by programs into addresses applied to the memory hardware. The 80286 can be set to operate in either "real address" mode (as used in the 8086) or "virtual address" mode—the setting of a bit in the status register selects the mode. In the latter mode the memory management characteristic of the 80286 uses the segment registers (CS, DS, SS,

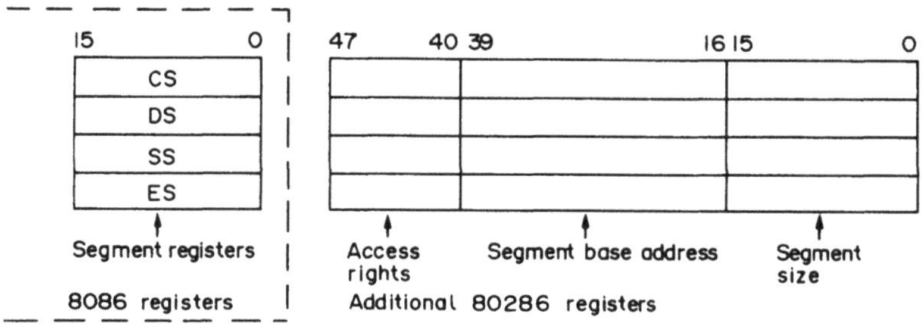

FIG. 6.11. 80286 segment descriptor cache registers.

ES—see section 6.3) in a different way to that of the 8086. The contents of these registers are not interpreted by the CPU as 16-bit components of the 20-bit physical address (see Fig. 6.4), instead they are treated by the CPU as memory pointers to 48-bit segment extension registers, as shown in Fig. 6.11. When a task (program) change instruction is executed, the CPU uses the contents of the segment registers to point to memory locations which contain the 48-bit segment descriptor registers. These 48-bit values are automatically transferred to the "cache" (cache means on-chip memory) registers shown in the diagram. The program is then implemented using these 48-bit registers to provide physical memory addressing; note the "access rights", which provides features such as write-protect. If the address specified is on disc (virtual memory), then the following procedure is followed:

(a) program requests access to segment currently on disc
(b) CPU checks descriptor table (transferred to cache on CPU)
(c) segment not present causes an interrupt
(d) operating system triggers CPU to transfer segment from disc
(e) CPU performs DMA transfer from disc
(f) operating system changes descriptor table
(g) operating system returns to trapped instruction.

A protection system is implicit when the 80286 is in virtual addressing mode, such that one program cannot modify another program, e.g. the operating system. This protection is provided by the 80286 system's four privilege levels. For example, an application program (running in the lowest privilege level) can execute a subroutine CALL instruction into the kernel of the operating system (set at the highest privilege level), but protection is provided to ensure that the operating system, although resident in the same sector of memory, cannot be over-written.

Intel offer a floating-point processor (the 80287) to support the 80286. There is no doubt that the operation of an 80286 in virtual address mode, and supported by an 80287, places such a microcomputer system into the same performance category as most 16-bit minicomputers, and possibly even beyond.

6.10 The 80130 OSP (Operating System Firmware Processor)

This component is offered by Intel to act as a co-processor to either an 8088 or 8086 CPU. It executes 35 operating system primitive instructions, i.e. it effectively extends the 8088/8086 instruction set. The operating system is therefore partially provided in hardware, thus increasing the execution speed compared with a full software implementation. The 80130 possesses internal features such as programmable interrupt logic, timers and baud rate generator to assist its role of performing part of the operating system's task (e.g. job management, interrupt handling, memory management).

Bibliography

1. *16-bit Microprocessor Handbook*. Adam Osborne and Gerry Kane. Osborne/McGraw-Hill, 1981.
2. *16-bit Modern Microcomputers—the Intel I8086 Family*. George W. Gorsline. Prentice-Hall, 1985.
3. *iAPX 88 Book*. Intel, 1981.
4. *The 80286 Architecture*. Stephen P. Morse and Douglas J. Albert. Wiley, 1986.
5. *8086/8088 16-bit Microprocessor Primer*. Christopher L. Morgan and Mitchell Waite. McGraw-Hill, 1982.

CHAPTER 7

MOTOROLA MC68000 FAMILY (16-BIT)

7.1 Introduction

Motorola introduced the 16-bit MC68000 family of microprocessors to succeed the MC6809 8-bit microprocessor, but the devices are not program compatible. The MC68000 series is arguably 32-bit, because the devices possess 32-bit work registers. There are three main processors in the MC68000 series:

(a) MC68000—24-bit address bus and 16-bit data bus; used in the Apple Macintosh computer
(b) MC68008—20-bit address bus and 8-bit data bus; used in the Sinclair QL computer
(c) MC68010—as MC68000 with "virtual machine/memory" characteristic (i.e. extends physical main memory to include backing store); used in the Hewlett Packard 9000 computer.

7.2 MC68000 Pin Functions

The pin functions of the MC68000 are shown in Fig. 7.1. The use of a 64-pin DIL package enables the address and data buses to be non-multiplexed. Additionally a large number of control bus signals can be accommodated. Therefore the 8288 bus controller and de-multiplexing chips required with the Intel 8086/8088 are unnecessary with the Motorola device. Additionally the MC68000 does not possess a "maximum/minimum" option which is selected by the setting of a control pin as with the Intel devices; the 64-pin package enables the "maximum" configuration to be permanently selected.

Pins UDS and LDS indicate whether data are being transferred on the most significant (upper) byte, least significant (lower) byte or both bytes of the data bus. FC0, FC1 and FC2 represent the function code (type of bus activity) of the device, e.g. user program access, supervisor data memory access, etc. Seven interrupt levels are coded on IPL0, IPL1 and IPL2. Three bus arbitration signals BR, BG and BGACK are employed when other processors or DMA controllers require to utilise the MC68000's buses. Three further signals E, VPA and VMA are provided to enable interfacing

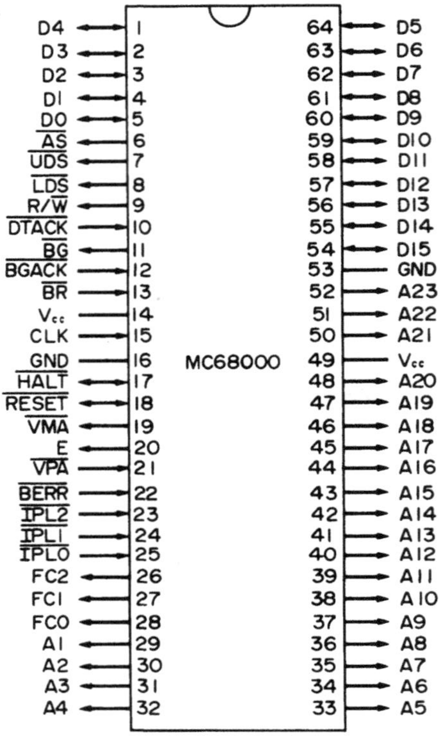

FIG. 7.1. MC68000 pin functions.

to standard Motorola MC6800 devices, e.g. the MC6821 PIO (or PIA)—the E signal is a synchronising clock signal which synchronises data transfers.

7.3 Processor Architecture

The work register block for the MC68000 series is shown in Fig. 7.2. There are seventeen 32-bit data and address registers, and a 32-bit program counter (only 24 bits are used). The eight data registers can handle data values as bytes, words (16 bits) or long words (32 bits). All of the registers can function as index registers.

Although 32-bit registers are applied, the ALU is only 16 bits wide. The MC68000 utilises a "pipeline" system in a similar but more limited way to the 8086/8088; it fetches an instruction code during the execution of the two prior instructions. The 16M bytes of memory are effectively expanded to 64M bytes by utilising the control bus lines FC0, FC1 and FC2.

The status register is detailed in Fig. 7.3, and is divided into two bytes—the system byte and user byte. The user byte contains the normal CPU status

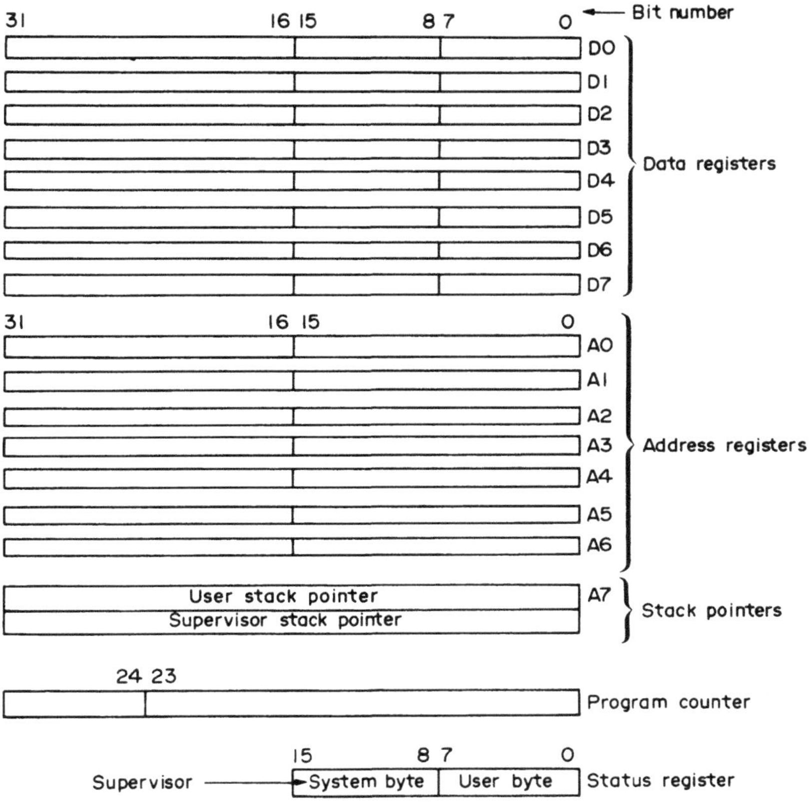

FIG. 7.2. Work registers for MC68000.

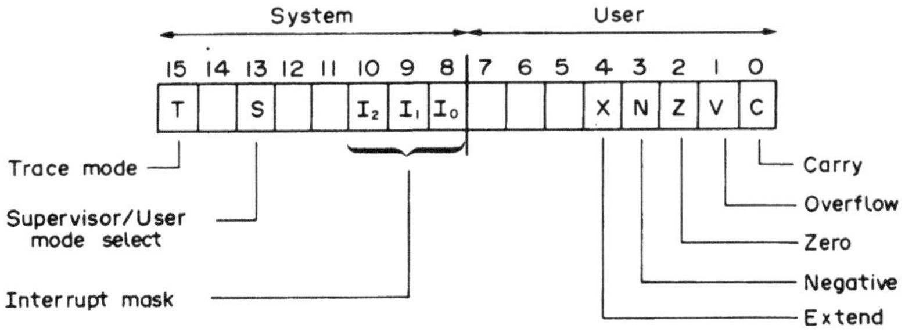

FIG. 7.3. Status register.

indicator flags (the Extend bit is used in multiprecision arithmetic operations), but the device can operate in the "supervisor" or "system" mode as selected by the Supervisor status bit. In supervisor mode certain privileged instructions can be used and a separate stack pointer is applied. There is no similar system with the Intel 8086/8088. The "trace" mode can be entered when in supervisor mode if the trace bit is set, such that after each instruction is executed a trap is forced. In this way a debugging program can monitor the results of the execution of that instruction.

Motorola call their interrupt system the "exception" system. There is some justification for this nomenclature because the MC68000 exception system handles a much broader range of events than the normal interrupt system in a typical microprocessor. Exceptions can be generated by external events or internal events. Externally generated exceptions are:

(a) Interrupts (via IPL0, IPL1 and IPL2)
(b) Reset (the RESET signal)
(c) Bus errors.

Internally generated interrupts are:

(a) Instruction traps
(b) Trace (as described above)
(c) Privilege violations (when an instruction that is reserved for supervisor mode is used in user mode)
(d) Illegal opcodes
(e) Addressing errors (when an odd address is specified—all memory accesses must be on even address boundaries).

The MC68000 processes 256 exceptions, and the start addresses of the programs for these 256 events are contained in the Exception Vector Table, which is shown in Fig. 7.4. The table occupies the first 1024 bytes in the memory address range, and each 4 bytes contain the 32-bit address (called "vector") which is loaded into the program counter as part of the exception processing sequence. The only variation is that the Reset exception uses two vectors—one to load the program counter and the other to load the supervisor stack pointer. Notice that vectors 64 to 255 allow for the interrupting device (several of which may be daisy-chained on one of the seven interrupt levels) to generate part of the exception vector, e.g. the interrupting device places 8 bits on the data bus lines D7 to D0. The sequence of operations that occurs when an external interrupt signal is set (assuming that its priority is higher than that contained in the interrupt mask) is:

(1) The contents of the status register are stored in an internal register,

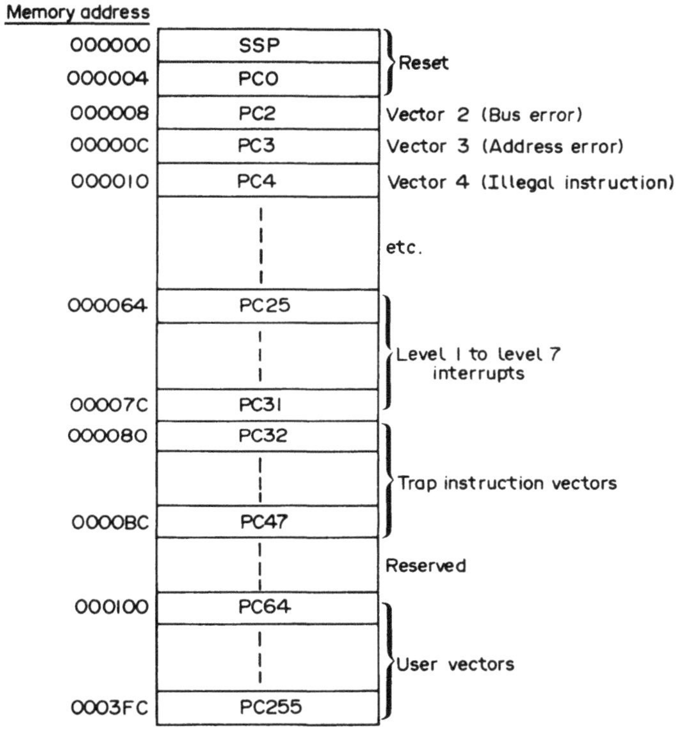

FIG. 7.4. Exception vector table.

supervisor mode is entered (S bit is set) and trace is disabled (T bit is cleared)

(2) The interrupt mask in the status register is updated with the new priority setting

(3) The exception vector byte is read in (on D7 to D0) from the interrupting device—this byte is multiplied by 4 to generate the full interrupt vector

(4) The program counter and status register are stored on the supervisor stack.

(5) The program counter is loaded with the four bytes from the appropriate location in the exception vector table, and instruction execution commences at the start address of the interrupt servicing program.

7.4 Instruction Set

Within each instruction the opcode is word (16 bits) length; 3 bits within the 16-bit word are used to specify register number (when relevant) and a

TABLE 7.1 *MC68000 Instruction Set*

Mnemonic		Description
Data move		
MOVE	source,dest	Move data value (byte, word or double-word) from source to destination (all addressing modes are allowed)
MOVE	register,dest	Move multiple registers (the contents of the specified registers are transferred to contiguous memory locations)
MOVEP	data reg,addr reg,displ	Move peripheral data (transfer contents of data register to address computed by adding displacement to contents of address register)
EXG	reg,reg	Exchange the contents of the specified registers
LEA	source,addr reg	Load source address into address register
SWAP	data reg	Exchange the 16-bit halves of a data register
PEA	address	Push effective address
Data modify		
ABCD	source,dest	Add decimal (BCD) with extend bit
ADD	source,dest	Add binary
AND	source,data reg	AND
ASL	data reg	Arithmetic shift left (variable shift count)
ASR	data reg	Arithmetic shift right (variable shift count)
BCHG	bit no,dest	Test a bit and change
BCLR	bit no,dest	Test a bit and clear
BSET	bit no,dest	Test a bit and set
BTST	bit no,dest	Test a bit
CLR	dest	Clear an operand (register or memory)
CMP	source,data reg	Compare (and set flags)
DIVS	source,data reg	Signed divide (16-bit)
DIVU	source,data reg	Unsigned divide (16-bit)
EOR	data reg,dest	Exclusive-OR
EXT	data reg	Sign extend data register (byte to word to double-word)
LSL	data reg	Logical shift left
LSR	data reg	Logical shift right
MULS	source,data reg	Signed multiply (32-bit result)
MULU	source,data reg	Unsigned multiply (32-bit result)
NBCD	dest	Negate decimal with extend
NEG	dest	Negate (two's complement)
NOT	dest	Logical complement (one's complement)
OR	source,data reg	OR
ROL	dest	Rotate left without extend (variable rotate count)
ROR	dest	Rotate right without extend
ROXL	dest	Rotate left with extend
ROXR	dest	Rotate right with extend
SBCD	data reg, data reg	Subtract decimal with extend bit
SUB	source,dest	Subtract binary
Branch/jump		
Bcc	displ	Branch conditionally by displacement relative to PC (with variety of conditions, e.g. BLE −24)
BRA	displ	Branch always (relative to PC)
BSR	displ	Branch to subroutine
DBcc	data reg,displ	Test condition, decrement and branch
JMP	address	Jump to address
JSR	address	Jump to subroutine

TABLE 7.1 *Continued*

Mnemonic		Description
RTE		Return from exception
RTR		Return and restore (status register)
RTS		Return from subroutine
Control/miscellaneous		
CHK	source,data reg	Check register against bounds (and generate exception)
LINK	address reg, displ	Link stack and allocate (address register to stack)
NOP		No operation
RESET		Reset external devices
Scc	address	Set (byte) according to condition
STOP		Stop
TAS	dest	Test and set an operand (to synchronise other CPUs)
TRAP	vector	Trap
TRAPV		Trap on overflow
TST	dest	Test an operand (set flags)
UNLK	addr reg	Unlink (load SP from address register, and then address register from stack)

further 3 bits specify the addressing mode (when relevant). There are only 56 mnemonics (Table 7.1) in the instruction set, and this makes the instruction set an attractively simple one to master. There are no input/output instructions so input/output must be memory mapped. Both signed and unsigned multiply and divide instructions are available.

The length of an instruction varies from 1 to 5 words (2 to 10 bytes) depending on which addressing mode is being used and which data length is required (byte, word or long word). The 14 different addressing modes can be grouped into 6 basic types, as follows:

(a) Direct register
`MOVE.W A4,D2` ;Load D2 with contents of register A4
(Note: W denotes Word, B denotes Byte, L denotes Long Word)

(b) Direct memory
`MOVE.W $500200,D2` ;Load D2 with contents of memory location hex. 500200

(c) Indirect
`MOVE.W (A1),D2` ;Load D2 with contents of memory address held in A1

Possible variations of this mode are auto increment/decrement, indexing and displacement.

(d) Immediate
`MOVE.W #3,D2` ;Load D2 with 3 (using 16-bit word length)

(e) Implied

```
PEA A6                  ;Push effective address (contents of A6) to stack
```

(f) Program counter relative

```
BNE $Ø7                 ;Branch (jump) if not equal forward 7 bytes
```

Once again indexing and displacement can be applied to this mode. Additionally absolute address jump addressing is available, e.g.

```
JMP $2ØØØØØ             ;Jump to address hex. 200000
```

Notice that the source and destination are specified in this assembly language mnemonic form in the reverse order to that applied for the Intel 8088/8086 instructions.

The following program examples demonstrate use of the instruction set.

Program Example 1

```
MOVE.W #5,D3          ;Move 5 (16-bit word) into data register 3
ADD.W  $1BØØ43,D3     ;Add word from memory address hex. 1B0043
MOVE.B D3,$2ØØØØØ     ;Output lower byte to PIO
```

This program uses the simplified circuit configuration of Fig. 7.5, which

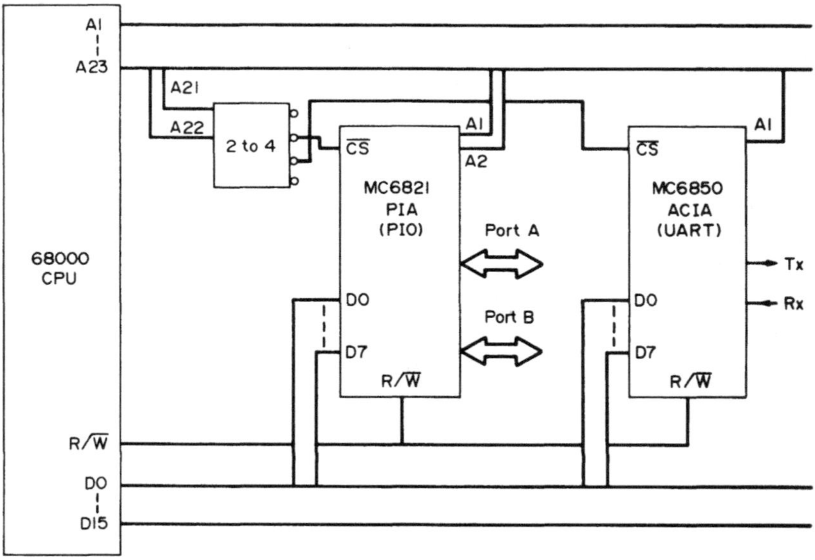

FIG. 7.5. Simplified MC68000 input/output circuit.

utilises a standard Motorola MC6821 PIO (called PIA—Peripheral Interface Adaptor) and illustrates the manner in which 8-bit input/output devices are applied with the 16-bit MC68000 processor. The program section adds two bytes and outputs the result to an output port.

Program Example 2

;This program (using circuit of Fig. 7.5) initialises a UART to required baud
;rate, number of data bits and parity, then outputs 10 "Line Feed" ASCII
;characters to the transmit pin Tx. The UART (MC6850 ACIA—
;Asynchronous Communications Interface Adaptor) has 2 addresses—
;hex. 400000 (control register output and status register input registers) and
;hex. 400002 (Tx output and Rx input data transfer registers).

```
PROGRAM   EQU     $3000      ;Start address of program
UARTCON   EQU     $400000    ;Address of UART control and status
                              registers
UARTTX    EQU     $400002    ;Address of UART Tx (transmit)
                              register
RESET     EQU     $03        ;Control word for master reset
INIT      EQU     $49        ;Control word for 4800 baud, 7 data bits,
                              even parity and 1 stop bit
NUMBER    EQU     9          ;Number of characters
LINEFEED  EQU     $0A        ;ASCII for Line Feed
BUSYBIT   EQU     $1         ;Status register transmit busy bit
          ORG     PROGRAM
          MOVE.B  #RESET,UARTCON      ;Reset UART (control
                                       register)
          MOVE.B  #INIT,UARTCON       ;Initialise UART
          MOVE.B  #NUMBER,D1          ;Place number of charac-
                                       ters in D1
TRANSMIT  BTST.B  #BUSYBIT,UARTCON    ;Test if UART busy
          BEQ.S   TRANSMIT            ;Poll until not busy
          MOVE.B  #LINEFEED,UARTTX    ;Output Line Feed to Tx
          DBRA    D1,TRANSMIT         ;Repeat loop until 10
                                       operations
          END
```

Notice that the last instruction (DBRA) decrements the data register D1 and branches until the result in D1 is −1. The # symbol denotes that immediate addressing is used.

Bibliography

1. *M68000 16/32-bit Microprocessor Programmer's Manual.* Motorola. Prentice-Hall, 1984.

2. *Using and Troubleshooting the MC68000.* James W. Coffron. Reston, 1983.
3. *68000 Microprocessor Handbook.* Gerry Kane, Osborne/McGraw-Hill, 1981.
4. *68000 Assembly Language Programming.* Gerry Kane, Doug Hawkins and Lance Leventhal. Osborne/McGraw-Hill, 1981.
5. *16-bit Microprocessors.* Ian R. Whitworth. Granada, 1984.

CHAPTER 8

ZILOG Z8000 FAMILY (16-BIT)

8.1 Introduction

The Zilog Z8000 series of 16-bit devices includes several CPUs, with differing capacities for memory support (including virtual memory), and several support devices. These support devices include powerful input/output handling chips and co-processors, e.g. floating point hardware chip.

The two principal CPUs, which are almost identical except for memory addressing capacity, are:

(a) Z8001—8M bytes memory (16 + 7 address lines) and 16-bit data bus in a 48-pin DIL package
(b) Z8002—64K bytes memory (16 address lines) and 16-bit data bus, in a 40-pin DIL package.

The Z8010 memory management unit is used in conjunction with the Z8001 in order to manage the 8M byte address space (providing segment relocation as well as memory protection), whilst the Z8003 and Z8004 are CPUs that are virtually identical to the Z8001 except that they offer the facility of virtual memory.

8.2 Z8001 Pin Functions

Figure 8.1 shows the pin functions of the Z8001; the 40-pin Z8002 does not possess the segment number signals (SNO to SN6) and the segment trap signal ($\overline{\text{SEGT}}$). Each Z8000 CPU generates the "Z-bus", which consists of address, data and control signals which enable data transfers between CPU and memory or input/output. All Z8000 CPUs possess multiplexed address and data pins—the setting of the signals $\overline{\text{AS}}$ (Address Strobe) indicates that address information is present on these lines. Signals SN0 to SN6 on the Z8001 are memory segment number signals and they act as additional address lines—increasing binary counts on these signals switch to different 64K "segments". The four status signals ST0 to ST3 can be decoded to produce 16 discrete signals to assist in allocating separate memory spaces for programs, data and stack. This is particularly useful with the Z8002 because

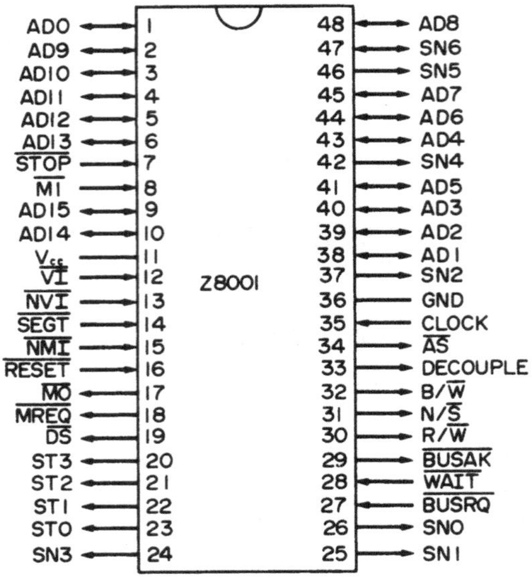

FIG. 8.1. Z8001 pin functions.

it allows the 64K addressing range to be expanded. Notice the following additional signals:

(a) apart from $\overline{\text{RESET}}$ there are four interrupt signals: $\overline{\text{NMI}}$ (non-maskable interrupt), $\overline{\text{NVI}}$ (non-vectored interrupt), $\overline{\text{VI}}$ (vectored interrupt) and $\overline{\text{SEGT}}$ (segmentation trap—used when the memory management unit is used with the Z8001)
(b) DMA signals $\overline{\text{BUSREQ}}$ and $\overline{\text{BUSAK}}$
(c) status signals B/$\overline{\text{W}}$ (byte/word), N/$\overline{\text{S}}$ (normal/system mode) and R/$\overline{\text{W}}$ (read/write)
(d) daisy-chained multi-micro control signals $\overline{\text{MI}}$ and $\overline{\text{MO}}$ to allow one CPU to access a shared device in a multi-processor system.

8.3 Processor Architecture

The work register block for the Z8001 is shown in Fig. 8.2. There are sixteen 16-bit registers, all of which can be used as accumulators and all except R0 can be used as index registers. R0 to R7 can process data in byte form as well as word form, whilst the sixteen registers can be used in pairs to offer 32-bit working. It is also possible to perform some 64-bit (4 word) working. The program counter consists of a 16-bit "offset" and a 7-bit segment number. The CPU can be operated in "system" or "normal" mode (analogous to the supervisor and user modes with the Motorola MC68000),

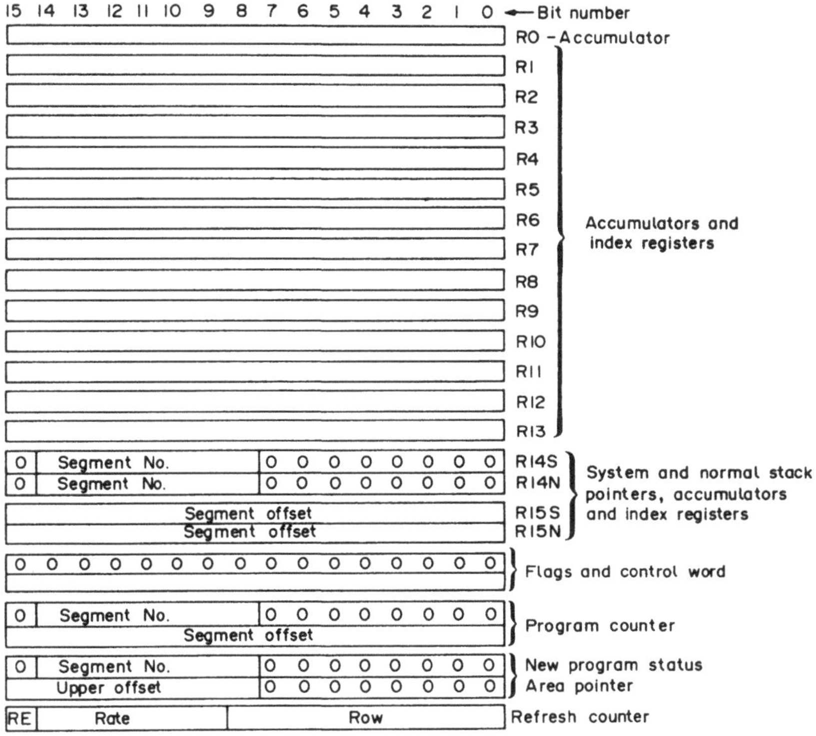

FIG. 8.2. Work registers for Z8001.

and a bit in the status register marks the current mode. There are two stack pointers, one for each mode, and one of the two R15 registers (plus 7 bits in each R14) is used to represent the stack pointer (S = System, N = Normal). Additional instructions are available when in system mode (as with the MC68000). In particular the input and output instructions fall into this category, so that in a multi-programming application a user program (operating in normal mode) must demand that the operating system (operating in system mode) performs input/output tasks. Zilog have maintained the use of a refresh counter (to refresh dynamic RAM) in line with the operation of the 8-bit Z80.

There is no pipeline system with an on-chip instruction queue, as employed by the Intel 8086/8088, but the Z8000 CPUs employ an early instruction decode system. This allows decoding to occur independently of the addressing mode chosen.

Memory addressing is arranged in bytes, but A0 is utilised as with the Intel 8086, i.e. to select one 8-bit bank in lower (even) or upper (odd) address banks. Each opcode is word length, and opcode and data words always start at an even memory address.

8.4 Instruction Set

There are nine addressing modes, as follows:

(a) Register

```
LD R4,R7              ;Load R4 with content of R7 (register 7)
```

(b) Immediate

```
LD R4,5               ;Load 16-bit word 5 into R4
```

(c) Indirect register

```
LD R4,(R2)            ;Load R4 with contents of memory location
                       held in R2
```

(d) Direct address

```
LD R4,%1800           ;Load R4 with contents of memory location
                       hex. 1800
```

(e) Index

```
LD R4,%4000(R1)       ;Load R4 with contents of memory location
                       computed by adding hex. 4000 to the con-
                       tents of R1
```

(f) Base

```
LD R4,(R6)(4)         ;Load R4 with contents of memory location
                       computed by adding the displacement 4 to
                       the contents of R6
```

(g) Base indexed

```
LD R4,(R5)(R12)       ;Load R4 with contents of memory location
                       computed by adding the index value in R12
                       to the base address in R5
```

Note: In all these examples, except (b), the source and destination can be interchanged, of course.

(h) Relative

```
JR +17                ;Jump (unconditional) to PC+17
```

Normal absolute address jump/branch addressing mode is available, e.g.

```
JP %0400              ;Jump (unconditional) to hex. 0400
```

(i) Implied

```
LDCTLB FLGR,%31       ;Load 31 into control byte register
```

The instruction set is summarised in Table 8.1, and it uses similar mnemonics to the Z80, as the following Z8001 program example shows:

Program Example

```
LD   R1,#%643D        !Load 643D into R1!
LD   R2,<<3>>%5000    !Load R2 with contents of memory
                       location offset 5000, segment 3!
ADD  R2,R1            !Add R1 to R2!
OUT  3,RL2            !Output lower half of R2 to I/O port
                       address 3!
```

This program adds a constant to the contents of a memory location, and sends one half of the result to an output port.

TABLE 8.1 *Z8000 Instruction Set*

Mnemonic		Description
Data move		
LD	dest,source	Move data value (byte, word or long word) from source to destination—various addressing modes
LDM	reg,source,n	Load multiple (consecutive words) from memory to registers
LDM	dest,eg,n	Load multiple (consecutive words) from registers to memory
EX	reg,source	Exchange words (EXB—exchange bytes)
IN	reg,source	Input to register from source (INB—input byte)
OUT	dest,reg	Output from register to source (OUTB—output byte)
Additional range of input/output instructions, e.g.		
INIR	dest,source,reg	Input from source to destination, autoincrement destination address, decrement register and repeat until register contents are zero
PUSH	index reg, source	Push
POP	dest, index reg	Pop
Data modify		
ADC	reg,source	Add with carry
ADD	reg,source	Add (ADDB—add bytes; ADDL—add long words)
CP	reg,source	Compare
DAB	dest	Decimal adjust
DEC	dest,n	Decrement by n
DIV	reg,source	Divide (signed)
EXTS	dest	Extend sign (from lower half through to higher half)
INC	dest,n	Increment by n
MULT	reg,source	Multiply (signed)
NEG	dest	Negate
SBC	reg,source	Subtract with carry
SUB	reg,source	Subtract
AND	reg,source	AND
COM	dest	Complement
OR	reg,source	OR
TCC	cond,dest	Test condition code (set LSB)

continued

TABLE 8.1 *Continued*

Mnemonic		Description
TEST	dest	Test
XOR	reg,source	Exclusive-OR
BIT	dest,b	Test bit static (b can be replaced by register)
SET	dest,b	Set bit static
TSET	dest	Test and set
RL	dest,n	Rotate left (RLB—rotate byte)
RLC	dest,n	Rotate left through carry
RLDB	reg,source	Rotate digit left
RR	dest,n	Rotate right
RRC	dest,n	Rotate right through carry
RRDB	reg,source	Rotate digit right
SDA	dest,reg	Shift dynamic arithmetic (by contents of register)
SDL	dest,reg	Shift dynamic logical
SLA	dest,n	Shift left arithmetic
SLL	dest,n	Shift left logical
SRA	dest,n	Shift right arithmetic
SRL	dest,n	Shift right logical
Branch/jump		
CALL	dest	Call subroutine
CALR	dest	Call relative
DJNZ	reg,dest	Decrement register and jump if non-zero
IRET		Interrupt return
JP	cond,dest	Jump conditional
JR	cond,dest	Jump conditional relative
RET	cond	Return conditional
SC	source	System call
Control/miscellaneous		
COMFLG	flags	Complement flag
DI		Disable interrupts
EI		Enable interrupts
HALT		Halt
LDCTL	CTLR,source	Load into control register
NOP		No operation
Range of multi-micro control operations, and block transfer and string manipulation instructions, e.g.		
CPSD	dest,source, reg,cond	Compare string and decrement (register)

8.5 Memory Management and Virtual Memory

Zilog offer an extremely useful device to support the Z8001 CPU in order to manage its 8M byte memory addressing space. This is the Z8010 memory management unit, as illustrated in Fig. 8.3. The device converts the 23-bit "logical" address set by the CPU into a 24-bit "physical" address that is applied to the memory circuit. A conversion is performed in the MMU (memory management unit), such that the seven segment registers (SN6 to SN0) are used in a look-up table to generate a 16-bit block identifier (each

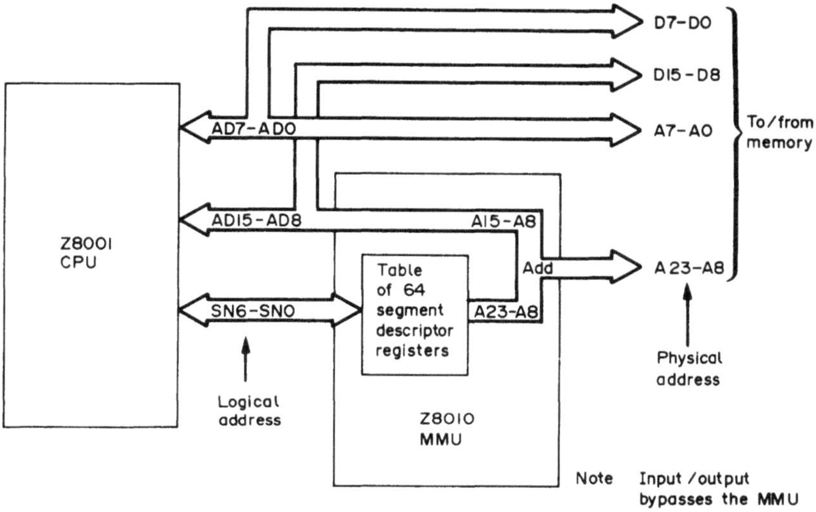

FIG. 8.3. Z8010 memory management unit.

block is 256 bytes). The low-order 8 address lines bypass the MMU. The physical addresses therefore can access 64 segments in memory, and each segment can have from 256 to 64K bytes. The value of this arrangement is:

(a) The amount of physical memory in a system may be smaller than the required logical memory for all the programs. The MMU can process two different programs with the same logical address ranges, and route them to different areas of physical memory. A program can therefore be relocated anywhere in physical memory by the MMU. An operating system can re-schedule the physical memory for a program to enable it to be run in a free area of memory in a multi-programming situation.
(b) Several attributes can be assigned to each segment within the MMU to provide memory protection, e.g. read only, DMA accesses inhibited, system mode only.

A memory management system gives provision for one more feature that is helpful in multi-programming situations. "Virtual memory" is an arrangement in which backing store is treated as if it is an extension of main memory; it is a common feature with mainframe computers and minicomputers. The operating system, memory management unit and disc combine to make physically addressable memory appear larger to users' programs. A flag can be set in the memory management unit for a particular program/segment such that the CPU is notified (by interrupt) and the MMU is re-programmed to transfer an additional segment from disc into memory. The MMU may have to make space in memory to accommodate this additional segment (by temporarily dumping another segment to disc). This memory manipulation is transparent to the user program. Zilog offer the Z8003 and Z8004 CPUs to replace the Z8001 and Z8002 respectively so that virtual memory operations can be implemented.

8.6 Supporting Chips

The Z80 8-bit support chips (PIO, SIO, CTC) can be utilised with Z8000 microprocessors if the address/data bus is de-multiplexed. However, Zilog offer a new range of supporting chips to their 16-bit CPUs. These devices are available in two series:

(a) Z-BUS peripherals, which connect to the Z8000 CPU buses, including the multiplexed address/data bus—identified by the number Z80XX
(b) Universal peripherals, which connect to conventional non-multiplexed non-Zilog CPUs—identified by the number Z85XX.

Most devices in these two ranges are compatible in function, as Table 8.2. shows.

Whilst the Z8X31 is basically a dual-channel UART, the Z8X30 offers asynchronous and synchronous modes as well as byte and bit protocols. The latter enables the bit-oriented protocols such as IBM SDLC (synchronous data link control) and HDLC (high-level data link control), which are more sophisticated than the simple asynchronous RS232-C links, to be implemented. The Z8X36 offers 3 counter/timers (16-bit) and 2 ports, whilst the Z8X90 offers 2 counter/timers, 2 ports and interrupt handling facilities. The FIFO devices facilitate buffering between CPUs (or between a CPU and a peripheral device).

The Z8070 floating point unit (FPU) is an example of what Zilog term their "extended processing architecture"—the FPU is described as an EPU

TABLE 8.2 *Zilog Z8000 Peripherals*

Z-BUS	Universal	Function
Z8016	Z8516	DMA controller
Z8030	Z8530	Serial communications controller
Z8031	Z8531	Asynchronous serial communications controller
Z8036	Z8536	Counter/timer and parallel I/O unit
Z8038	Z8538	FIFO input/output interface unit
Z8060	Z8560	FIFO buffer unit and Z-FIO expander
Z8068	—	Data ciphering processor
Z8070	—	Floating point unit
Z8090/4	Z8590/4	Universal peripheral controller

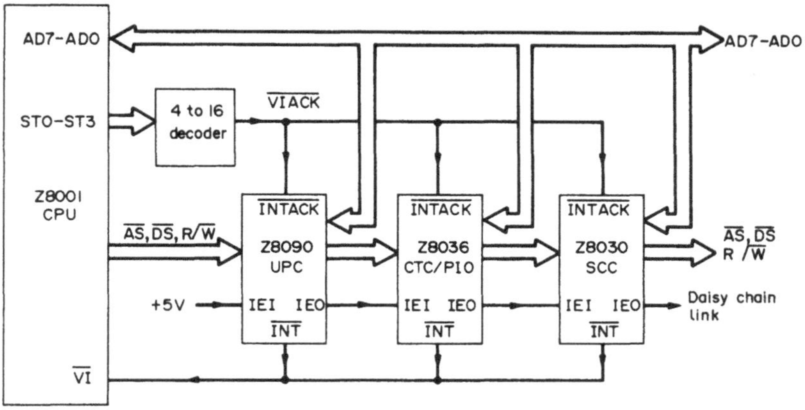

FIG. 8.4. Zilog Z8000 input/output interrupt system (simplified).

(extended processing unit). It connects to the CPU buses and performs floating point arithmetic operations while operating in parallel with the CPU. It monitors the same instruction stream as the CPU, and identifies and executes those instructions intended for it. Although the FPU operates internally using an 80-bit floating point format, data transfer between registers in the FPU and CPU is in byte or 16-bit form.

Zilog employ their standard "daisy-chain" system to enable several peripheral devices to share the same interrupt signal, as shown in Fig. 8.4. In this circuit example three standard Z-BUS peripheral devices are included, and each receives the least-significant half of the multiplexed address/data bus. The devices share the same interrupt signal VI (vectored interrupt), such that each device must generate its own identifying 8-bit code, which is used to establish the interrupt vector, when it requires to interrupt the CPU. The hardware priority system is established using the IEI (interrupt enable in) and IEO (interrupt enable out) daisy-chain. This interrupt system is unnecessary, of course, if the devices are to be polled only.

Bibliography

1. *Zilog Components Data Book*. 1985.
2. *The Z8000 Microprocessor*. Bradly K. Fawcett. Prentice-Hall, 1984.
3. *16-bit Microprocessors*. Ian R. Whitworth. Granada, 1984.

CHAPTER 9

32-BIT MICROPROCESSORS

9.1 Introduction

Following the success of 16-bit microprocessors throughout the 1980s it was inevitable that the principal manufacturers should extend the evolutionary process into 32-bit devices. The principal advantages offered by 32-bit devices over their 16-bit predecessors are:

(a) 32-bit data manipulation—larger integer numbers
(b) larger memory addressing range—normally 4 gigabytes (2^{32})
(c) faster operation—clock speeds of 16 MHz or more
(d) additional instructions and addressing modes—upwards compatible with their predecessors
(e) intrinsic memory management features
(f) instruction cache—on-chip memory holding most frequently used instructions and data items
(g) approximately 2 to 3 times improvement in processing speed for standard programming benchmark tests.

Manufacturers claim that 32-bit microprocessors rival traditional minicomputers, e.g. the DEC VAX machines, in computing power. Although this is debatable (in terms of instruction speed, co-processor—e.g. floating-point—speed and complex operating system support), 32-bit microcomputer systems are achieving widespread application in the fields of engineering workstations, speech recognition, robotic systems, office automation and large multi-user and multi-processing situations.

The four 32-bit devices that have been developed, and are described in later sections, are:

(1) Intel 80386
(2) Motorola MC68020
(3) Zilog Z80000
(4) Inmos T424 transputer (plus several other devices).

Whilst the first three devices represent natural progressions from their 16-bit counterparts, and use conventional CPU architectures, the Inmos transputer is a completely novel approach to machine architecture. It is an example of a RISC (reduced instruction set computer) compared with the traditional CISC (complex instruction set computer), and is designed to operate in a multi-processor configuration, i.e. several transputer CPUs execute a program task in parallel. The concept of RISC represents an attempt to diverge from the evolutionary manner in which CPU design has become increasingly complex with increasing instruction set features. Several research organisations and universities have attempted to produce CPUs with far smaller numbers of instructions offering increased performance and speed of operation. The essential features of a true RISC processor are single-cycle operation (multiple memory transfers are to be avoided) and hardwired control (instruction execution is implemented using fast hardwired logic rather than microcode—microprocessors employ the slow "microcode" technique of a look-up process to determine CPU operations to implement each instruction). Only the Acorn ARM (Acorn RISC machine) and (arguably) the Inmos transputer are commercial 32-bit RISC processors currently available, but it is possible that future computer architectures might lean more to this approach to achieve higher performance.

The 80386, MC68020 and Z80000 32-bit processors employ instruction cache and memory management facilities, and both features deserve more detailed descriptions here. A cache memory is a high-speed memory that is either contained on the CPU itself, or is placed between the CPU and main memory. Large main memory systems are invariably dynamic RAM, which is cheap but slow compared with static RAM. If the most frequently addressed instructions and data are held in fast static RAM cache memory, program execution speeds can be enhanced. Most programs tend to reaccess the same memory addresses, and the cache memory holds the contents of these addresses, together with the address itself. When program execution demands the contents of one of these addresses, e.g. the reading of a program instruction, the cache performs a high-speed compare to determine if the "tag" address requested by the CPU matches one of the stored items within the cache. If it is, a "hit" occurs, and the instruction can be read from cache in place of being transferred from memory with the associated inherent time delay. The hit rate must be high (typically greater than 80%) to make the cache system worthwhile. Typical cache sizes are 4K bytes—the larger the cache size, the higher is the hit rate.

The second advanced feature that 32-bit processors employ is memory management. Memory management is applied to allocate different areas of memory to different programs (and data areas) as efficiently as possible, and also to provide access protection to these programs. Once again memory management can be built onto the CPU chip itself, or it may require an additional component. An MMU (memory management unit) translates the "logical" memory address generated by the CPU into a "physical" address that is applied to the memory circuit. Thus, for example, an operating system can transfer control from one program to another, whilst both programs may share the same logical address range but are located separately in physical memory. Additionally the MMU can provide protection of program or data, e.g. read-only and privilege levels.

32-bit microprocessors give provision for co-processors, the most common of which is a floating-point arithmetic processor. All such floating-point co-processors adhere to the IEEE P754 standard, viz. 80-bit double extended precision.

Fabrication technologies are either NMOS or CMOS, and each 32-bit microprocessor possesses between 200,000 and 300,000 transistors. Numbers of interconnection pins are far too great, e.g. 84-pin, to enable the conventional DIL package to be used, and so chip carrier packaging (four sides of interconnection pins) is used.

9.2 Intel 80386

The Intel 80386 (or 386) is a development of the 8088/8086 family of 16-bit processors. Therefore a software compatibility with the 8088, 8086 and 80286 is retained, which enables the wide software base available with the 16-bit family to be employed on 80386 systems. The 80386 applies on-chip memory management, but no on-chip instruction/data cache—this must be supplied by an external tag memory system.

The CPU architecture is similar to the 8088/8086 model, and Fig. 9.1

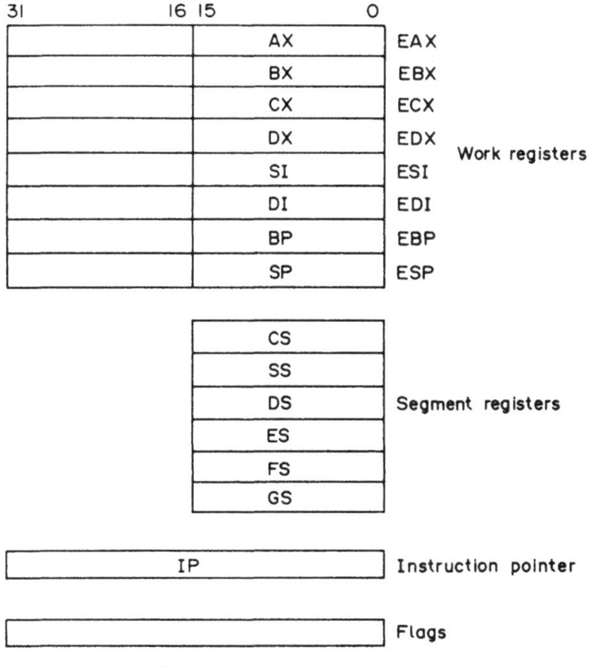

FIG. 9.1. 80386 register set.

TABLE 9.1 *Integer number range of 8-, 16- and 32-bit CPUs*

CPU	Integer range
8-bit	−128 to +127 (-2^7 to $+2^7 - 1$)
16-bit	−32,768 to +32,767 (-2^{15} to $+2^{15} - 1$)
32-bit	−2,147,483,648 to +2,147,483,647 (-2^{31} to $2^{31} - 1$)

illustrates the 80386 register set. There are eight 32-bit work registers, which can be used to perform calculations and to generate memory addresses. The six segment registers are applied to separate programs, data areas and the stack into different memory segments. Data types that can be processed are integer (two's complement—8-bit, 16-bit and 32-bit operations can be implemented), ordinal (unsigned integers—again 8-bit, 16-bit and 32-bit), BCD, string, bit and floating-point (if a floating-point co-processor is added). The exponential increase in integer number range achieved by a 32-bit processor compared with its 16-bit and 8-bit predecessors is illustrated in Table 9.1.

Some 64-bit operations, e.g. shift, are possible with the 80386. A wide range of addressing modes is available, including those offered on the 8088/8086, and several modes can be combined within a single instruction. An address can be computed within an instruction as follows:

(Base register) +	(Index register) *	(Scale) +	(Displacement)
↑	↑	↑	↑
Any register	Any register (except ESP)	1, 2, 4 or 8	Two's complement

The manner in which the 8086 uses its segment registers can be emulated by the 80386 if a bit in the Flags register is set. However the on-chip memory management features of the 80386 provide far more powerful operating system support using memory address translation and protection. The 80386 converts a logical address that is applied within a program instruction to a physical address, that appears on the pins of the device, using a system of on-chip tables. Firstly a base register is applied to select an entry in a "segment descriptor table" to generate a linear address, and secondly another base register is applied to select an entry in a "page table directory" to convert the linear address to the physical address. Information held in these tables also provides facilities for protection, e.g. read-only. To assist the page translation, the 80386 possesses an on-chip "translation lookaside buffer", which contains the 32 most recent conversions from linear address to physical address.

Both the 80287 and the 80387 floating-point co-processors can support the 80386. Additionally Intel have introduced the 82258 advanced DMA chip in order to transfer data directly between main memory RAM and peripherals, e.g. floppy disc, and the 82586 local area network co-processor. A typical 80386 system may therefore comprise:

(a) 80386 CPU
(b) 80287 or 80387 numeric co-processor (floating-point)
(c) memory cache system (e.g. 16K bytes static RAM)
(d) main memory (e.g. 4M bytes dynamic RAM)
(e) 82258 advanced DMA
(f) 8272 floppy disc control
(g) 82062 fixed disc control
(h) 82786 graphics co-processor
(i) 8259A interrupt controller
(j) 82530 serial control
(k) 82586 LAN control.

Backplane connections for multi-board systems follow the Intel "Multibus" standard.

9.3 Motorola MC68020

The architecture of the 32-bit Motorola MC68020 has evolved from the MC68000, MC68008 and MC68010 16-bit processor family. Machine code compatibility is retained, and the widespread software base that has been applied with MC68000 systems can be utilised on faster MC68020 configurations. The MC68020 offers more addressing modes, a small on-chip cache and external memory management support. It is mounted in a 114-pin grid array package.

Unlike the 16-bit family, the MC68020 possesses a small instruction cache and instruction pipeline system within the processor in order to speed up program execution times. The register set is similar, except that several additional registers are employed to handle the more powerful features that give multi-tasking operating system support. Figure 9.2 details the CPU register set, and the close similarity of this set to that of the MC68000 (see Fig. 7.2) is highlighted. Bit, BCD, byte, word (16-bits), long-word (32-bits) and quad word (64-bits) operations are available. The eight data registers act as accumulators, and the seven address registers can be used for base or indexed addressing. Only one of the three stack pointers can be active at one time—the setting of status bits in the status register determines which is active. The supervisor/user mode system is retained in the MC68020. For example, if the CPU is operating in supervisor (operating system) mode, as indicated by the setting of the appropriate status bits, and an interrupt occurs, the contents of the program counter, status register and exception vector offset are stored on the master stack automatically.

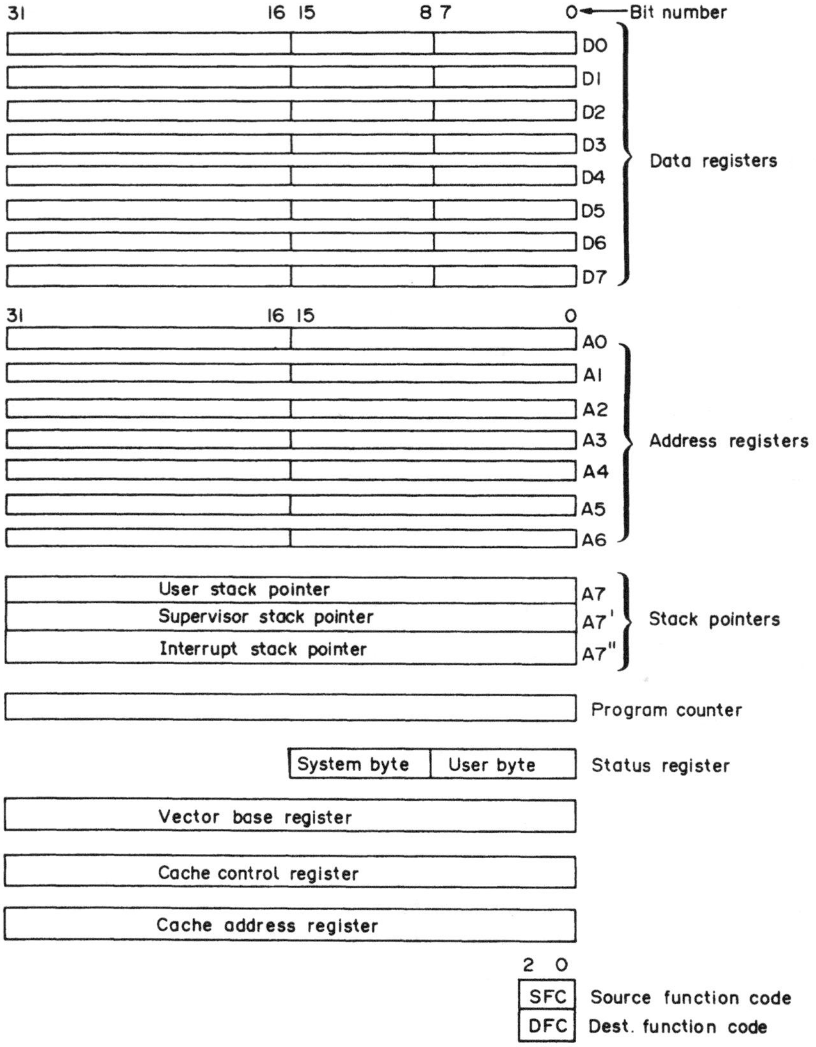

FIG. 9.2. MC68020 register set.

The vector base register is used to relocate the 1K byte exception vector table anywhere in memory. The two cache registers are used to control the instruction cache, e.g. enable the cache, clear the cache. The cache has 64 entries, and each entry contains one long word (64 bits), which includes a 32-bit tag field. The contents of one of the two function code registers appears on the interconnecting pins and are used to indicate one of the following: user program, user data, supervisor program, supervisor data.

A wide range of addressing modes is available, including register indirect with indexing and scaling, memory indirect with pre-indexing and post-indexing. As with its 16-bit predecessors, the MC68020 does not use input and output instructions, and so input/output must be memory mapped.

The MC68020 possesses the feature of "dynamic bus sizing". A 32-bit instruction is transferred from four consecutive memory byte locations, which must begin with an even-number address. However, data values can be transferred in byte, word (16-bits) or long word form, and the length of the data value is specified by the memory (or input/output) device itself. This is performed by the support device signalling to the CPU through two control signals (DSACK0 and DSACK1) the size of the data item. For example, whilst RAM is invariably 32 bits wide, e.g. using four 8-bit devices, it is sometimes easier to avoid placing a program or data list into four separate EPROMs. If the EPROM is to erased and re-blown, it may be preferable to place the whole program into one EPROM and transfer it one byte at a time when it is executed.

The small (64-entry) on-chip instruction cache of the MC68020 inevitably produces a low "hit" rate. It is advantageous therefore to augment this with an external static RAM cache memory, as applied with 80386 systems.

The MC68020 possesses co-processor control signals to enable interfacing between itself and the MC68881 floating-point processor. The normal IEEE P754 numeric operations, e.g. 80-bit floating-point operations, are available on this device. Motorola have also designed their MC68851 paged memory management unit to act as a co-processor. This unit provides address translation tables in memory. Additionally the memory management unit provides protection facilities, e.g. user access cannot be made to the operating system, user areas can be made read-only. A further feature included is that, if the CPU encounters a breakpoint instruction, the memory management unit provides the 32-bit instruction that is to be executed in its place. This is useful in program testing situations.

Multi-board configurations based on the MC68020 normally follow the Versabus or VME bus standards.

9.4 Zilog Z80000

The 32-bit Z80000 is, as expected, upwards-compatible with the 16-bit Z8000 processors (Z8001 and Z8002). In terms of its on-chip facilities it is probably the most advanced of the 32-bit microprocessors. It possesses an on-chip cache and on-chip memory management unit, and it will run at a clock speed of 25 MHz. A six-stage instruction pipeline is applied, and typical program execution speeds are 4 to 5 MIPS (million instructions per second).

The register set is illustrated in Fig. 9.3. As with the Intel and Motorola devices, data operations can be performed on bytes, words (16-bits), long

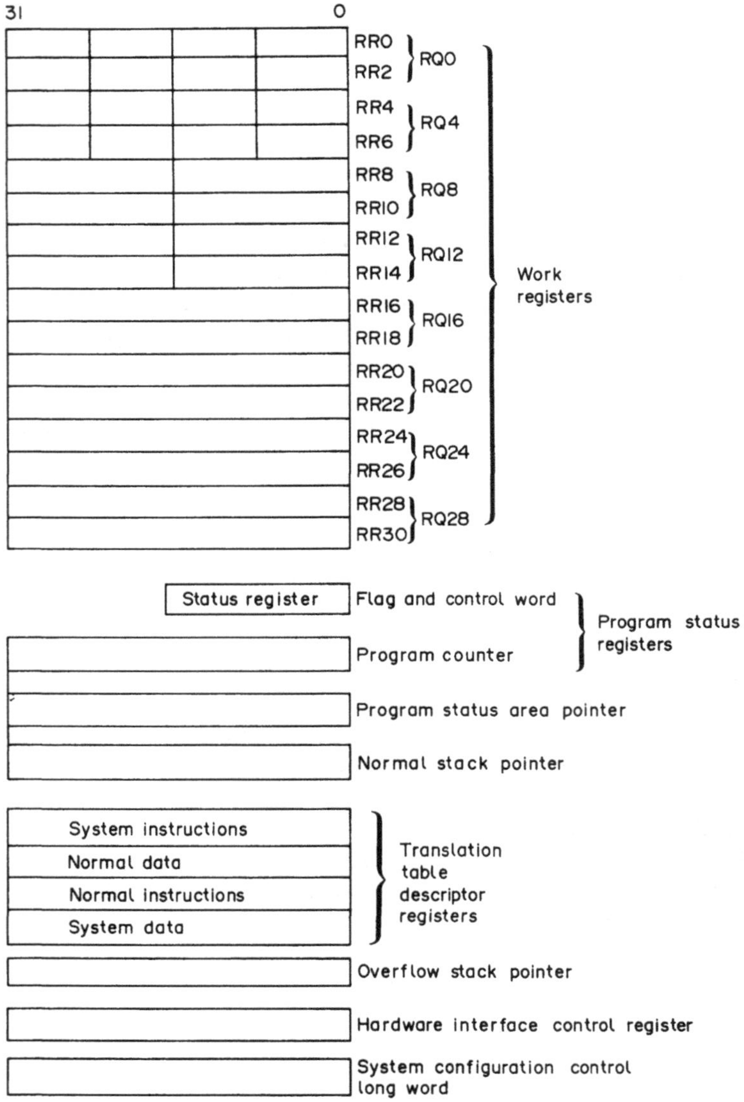

FIG. 9.3. Z80000 register set.

words (32-bits) and quad words (64-bits). One bit in the flag and control word (Zilog nomenclature for "status registers") indicates system or normal mode—some instructions, e.g. input/output, are only allowed when the CPU is set to system mode (when operating system is running). The contents of the program status register indicate the memory address of the values which are loaded into the program counter and flag and control word when an interrupt occurs. One of the four translation table descriptor registers is used by the memory management unit during memory addressing (to be described later). The overflow stack pointer is used if an address calculation error occurs when an interrupt is set. The system configuration control long word holds bits that control the on-chip cache and memory management functions.

The instruction set is comprehensive, and retains the Zilog features applied on earlier microprocessors, e.g. block transfer, string search. Nine addressing modes are available, and an address calculation can involve a base register, index register and a displacement. A memory address specified within an instruction is interpreted in one of three ways, as determined by the settings of two bits in the flag and control word. These three representations are:

(a) compact—16 bits (gives 64K bytes)
(b) segmented—32 bits (15-bit segment with 16-bit offset, for first 2G bytes)
(or, 7-bit segment with 24-bit offset, for next 2G bytes)

Note that only the offset field is affected by address calculations in this representation.

(c) linear—32 bits (gives 4G bytes)

This memory address specified in the instruction is termed the "logical address", and it is converted into the "physical address" in the memory management unit. The conversion process is performed in one of the following two ways:

(1) using the "translation lookaside buffer" within the CPU—this stores tag addresses and conversion information for the 16 most recently referenced pages in the same manner as an instruction/data memory cache
(2) if the translation lookaside buffer does not produce a tag match, the CPU references translation tables in memory, using one of the four table descriptor registers (see Fig. 9.3). and transfers the required information into the least recently used entry in the translation lookaside buffer

Additionally the memory management unit provides access protection, viz. execute, read and write.

The on-chip cache stores 16 entries of 16 bytes each. It maintains records of the 16 most recently addressed memory locations. If a cache miss occurs on an instruction fetch, "burst memory" transfers occur from consecutive memory locations to optimise the memory-read process.

As with the 16-bit Z8000 processors, the Z80000 refers to its co-processors as extended processing units (EPUs). Zilog apply the same floating-point co-processor—the Z8070—as used with their 16-bit processors. Other 16-bit support devices, e.g. the Z8016 DMA controller, can be applied with the Z80000. Zilog apply their own backplane interfacing standard ("Z-bus") to large microcomputer configurations.

9.5 Inmos Transputer

The Inmos transputer is an exciting and unique approach to computer design. Conventional computer architecture are based on a single CPU. Although the transputer is a single processor, it is not designed to operate in a uni-processor configuration, but instead is intended to operate in an arrangement in which several transputers operate in parallel. This multi-processor configuration produces a theoretical linear increase in performance as the number of processors increase, e.g. doubling the number of transputers doubles the number of instructions per second.

The name "transputer" is an amalgam of "transistor" and "computer". The execution of a program within several parallel processors is termed "concurrency". A large array of transputers is the first realistic solution to the much-vaunted Japanese concept of a fifth-generation computer, which must perform approximately 1000 million instructions per second.

The first transputer developed by Inmos was the T414, which is a 32-bit processor with 4 serial links and 2K bytes of on-chip static RAM. This was followed by:

(a) T424—32-bit, 4 serial links, 4K RAM
(b) T212—16-bit, 4 serial links, 2K RAM
(c) T800—32-bit, 4 serial links, on-chip floating point processor.

The T414 is fabricated using CMOS technology and is mounted in a 84-pin chip carrier package. At a clock speed of 5 MHz the device can perform 10 MIPS. Its internal organisation is shown in Fig. 9.4. In addition to a 32-bit CPU and 2K bytes of RAM, it includes four fast serial data links. These duplex links enable messages to be passed to other transputers which contribute to the concurrency of a programming task. Link adaptor chips are available to convert from serial to parallel data (bytes) to enable conventional input/output systems to be connected. Up to 4G bytes of memory can be addressed using the 32-bit address bus.

The processor module within the transputer is an example of a RISC machine (see section 9.1). It possesses an extremely simple instruction set,

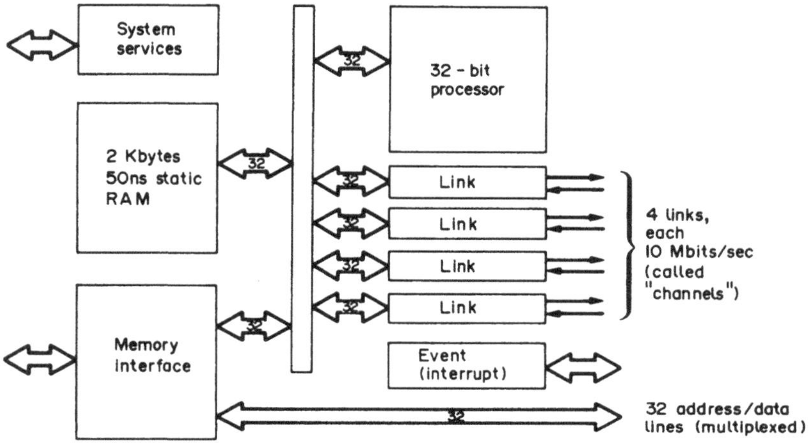

FIG. 9.4. Inmos T414 transputer architecture.

and most instructions execute in a single internal clock cycle. Figure 9.5 illustrates the restricted register set and simple byte instruction format.

It is not intended that the transputer is to be programmed at machine code level; indeed Inmos restrict information concerning the instruction set. A special high-level language named "occam" (after a fourteenth-century Oxford philosopher, who exhorted simplicity in problem-solving techniques) has been developed by Inmos to exploit the concurrency feature of transputer systems. An occam compiler is supplied by Inmos to produce highly efficient machine code for the transputer, and the programmer must specify which parts of the program are to run concurrently in other parallel transputers and which channels are to be used to communicate information. If a multi-transputer network is used, as shown in Fig. 9.6, different parts of the "process" (program) run in different transputers. The larger the array, the greater is the software throughput of the machine. Results quoted by

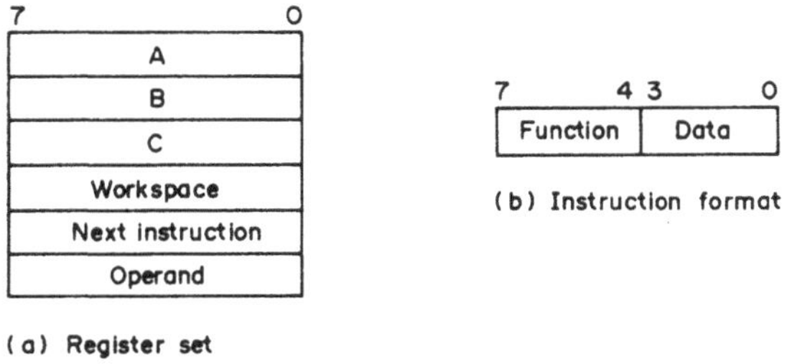

FIG. 9.5. Transputer CPU detail.

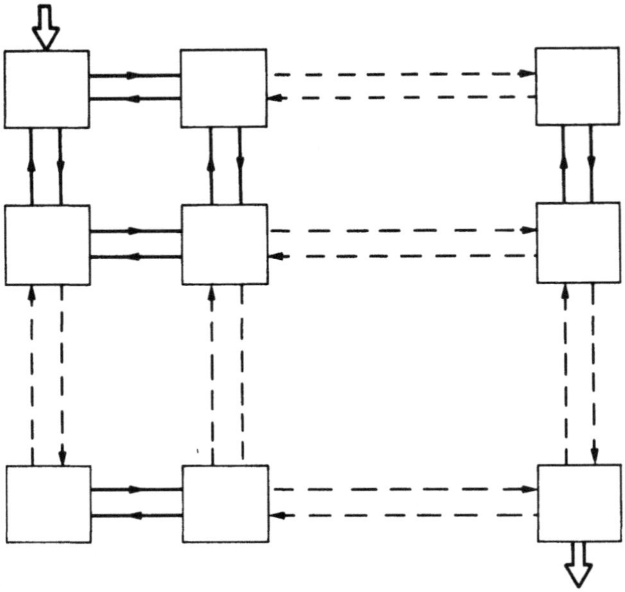

FIG. 9.6. Transputer array.

Inmos for large arrays are exciting, e.g. a mainframe performance exceeded by a 256-transputer system.

Support devices produced by Inmos include:

(1) F424 floating-point transputer
(2) G412 graphics controller transputer
(3) M212 disc controller transputer
(4) C001 and C002 link adaptors (to produce 8-bit parallel link).

Bibliography

1. *32-bit Microprocessors*. H. J. Mitchell, Collins, 1986.
2. *MC68020 32-bit Microprocessor User's Manual*. Prentice-Hall, 1985.
3. *Transputer Reference Manual*. Inmos, 1987.

CHAPTER 10

HIGH-LEVEL LANGUAGES

10.1 Summary of Programming Languages

The two programming language "levels" are:

(a) Low-level, i.e. a program expressed in machine code (normally written in hexadecimal bytes in the form in which it is stored in memory and executed by the CPU) and the more comprehensible assembly language equivalent (written using mnemonics for the instruction functions and labels to represent memory addresses).
(b) High-level, i.e. a program expressed in a form similar to spoken language (using recognisable words and arithmetic, and logical expressions to represent program functions).

The earlier chapters used low-level instructions and program examples to illustrate CPU operation. In this chapter examples of high-level language (HLL) programs are given, using the most commonly applied languages.

The obvious advantages of programming at high level are that programming is far easier and faster, and that programs are "transportable", i.e. they can be transferred to different computers with little or no alteration. Additionally the programmer does not need any knowledge of the CPU operation—he simply requires to follow the straightforward rules of writing program statements for the high-level language that he is using.

The procedure that a high-level language programmer normally follows is one of the following:

(1) He creates a text file, which is the source high-level language program, using an "editor" program. He then calls a "compiler" program, which converts his high-level language program into a machine code file—thus two files exist. When he runs his program, the machine code file is executed.
(2) He enters his text file into memory and calls an "interpreter" program, which converts each statement in his program into machine code and then executes it. No separate machine code file is created. Thus each time he wishes to execute his program, it must be converted into machine code. For this reason an interpreted program takes much longer to run than a compiled program, because the generation of machine code is performed as the program is executed.

High-level language interpreters are normally applied with single-user "home" computers, in which long program execution times are no hindrance. Compilers are used when programs must be executed quickly, e.g. in multi-user computers as applied in commercial and industrial applications.

The most common high-level languages are:

BASIC—has dominated the home microcomputer market (the only language that is normally interpreted, rather than compiled)
Pascal—becoming increasingly popular with microcomputers for educational applications due to its well-structured and modularized form
C—particularly popular with UNIX computer systems (see chapter 13)
FORTRAN—suitable for scientific programming
COBOL—principally used for commercial and business programming.

The dominant role of the first three languages in this list, particularly in educational establishments, justifies the following detailed descriptions.

10.2 BASIC

BASIC (Beginners' All-purpose Symbolic Instruction Code) is probably the simplest high-level language to learn, and for this reason it has become virtually an industry standard for small microcomputers. Its widespread application has unfortunately led to a diversity in its standardised form, and slight variations exist between different BASIC implementations.

The following are the principal statements that are applied in a BASIC program:

(a) LET—to set a numerical value to a variable name
(b) INPUT—to allow the operator to enter a number (or group of characters)
(c) PRINT—to display a number (or group of characters)
(d) GOTO—to transfer to another path within the program
(e) IF–THEN–GOTO—as for GOTO, but only if a specified condition is satisfied
(f) FOR–NEXT—to obey a section of program a specified number of times (a "loop")
(g) REM—a remark, or comment, which is not converted to machine code but serves simply to describe the action of the program
(h) DIM—to "dimension" a variable, i.e. specify the number of elements in an "array" of several values for the variable
(i) DATA–READ—the DATA statement establishes a list of numerical values, and the READ statement extracts one of these values (in order)
(j) GOSUB–RETURN—a subroutine is entered using the GOSUB statement; the subroutine terminates with the RETURN statement.

Consider the following sample program:

```
10  REM SAMPLE PROGRAM THAT INPUTS TWO NUMBERS
20  REM AND DISPLAYS THE SUM AND PRODUCT
30  PRINT "PLEASE ENTER TWO NUMBERS"
40  INPUT P,Q
50  REM NOW DISPLAYS ANSWERS
60  PRINT "SUM OF NUMBERS IS",P+Q
70  PRINT "PRODUCT OF NUMBERS IS",P*Q
80  END
```

When RUN is entered, the program produces the following display:

```
PLEASE ENTER TWO NUMBERS
?25,30 ←—— Operator enters 25 and 30
SUM OF NUMBERS IS   55
PRODUCT OF NUMBERS IS   750
```

The attractions of BASIC are:

(a) the program can easily be made to be "interactive" with the operator, e.g. the INPUT statement above causes program execution to stop at that point, a "?" is displayed and two numbers must be entered by the operator (separated by a comma, and terminated by RETURN)
(b) the use of line numbers enables the program to be altered simply, e.g. by re-entering a statement with a particular line number (such as 60), or a new line can be inserted by using an intermediate line number, before the program is run once more.

Most readers are probably familiar with the concepts of simple BASIC programs due to the widespread popularity of the language. The following program examples illustrate more detailed applications:

BASIC Program Example 1

PROGRAM DESCRIPTION (FOR STOCK CONTROL PROGRAM)

A newsagent's shop stocks ten different types of daily newspaper, and each newspaper is given a stock number from 1 to 10. At the start of every day differing numbers of newspapers are held for each type. The stock control program, written in BASIC, must:

(a) set up two arrays (A for stock number, B for number in stock) at the start of the day
(b) adjust the number in stock when one or more of each newspaper is sold during the day
(c) at the end of the day (when the operator enters a "rogue value" of −1) display the numbers held in stock.

PROGRAM LISTING

```
10   REM NEWSPAPER STOCK CONTROL PROGRAM
20   REM SET UP TWO ARRAYS
30   DIM A(10),B(10)
40   FOR I=1 TO 10
50   LET A(I)=I
60   PRINT "PLEASE ENTER NUMBER OF NEWSPAPERS FOR",I
70   INPUT B(I)
80   NEXT
90   REM NOW ENTER NEWSPAPER SALES
100  PRINT "PLEASE ENTER STOCK NUMBER FOR NEWSPAPER
     SOLD"
110  INPUT NUMBER
120  IF NUMBER=-1 THEN GOTO 240
130  REM CHECK IF THIS IS A VALID STOCK NUMBER
140  FOR J=1 TO 10
150  IF NUMBER=A(J) THEN GOTO 180
160  NEXT
170  PRINT "INCORRECT STOCK NUMBER";GOTO 100
180  PRINT "PLEASE ENTER NUMBER SOLD"
190  INPUT SOLD
200  IF SOLD<=B(J) THEN GOTO 220
210  PRINT "OUT OF RANGE NUMBER SOLD";GOTO 180
220  LET B(J)=B(J)-SOLD
230  GOTO 100
240  REM END OF DAY (-1 ENTERED)
250  REM NOW DISPLAY FINAL STOCK FIGURES
260  PRINT "END OF DAY STOCK FIGURES"
270  PRINT "STOCK NUMBER..NUMBER IN STOCK"
280  FOR K=1 TO 10
290  PRINT A(K),B(K)
300  NEXT
310  END
```

FLOW CHART

See Fig. 10.1.

TEST DATA

Typical test data that can be applied is shown in the following display that is obtained when the program is run:

```
PLEASE ENTER NUMBER OF NEWSPAPERS FOR    1
?40
PLEASE ENTER NUMBER OF NEWSPAPERS FOR    2
```

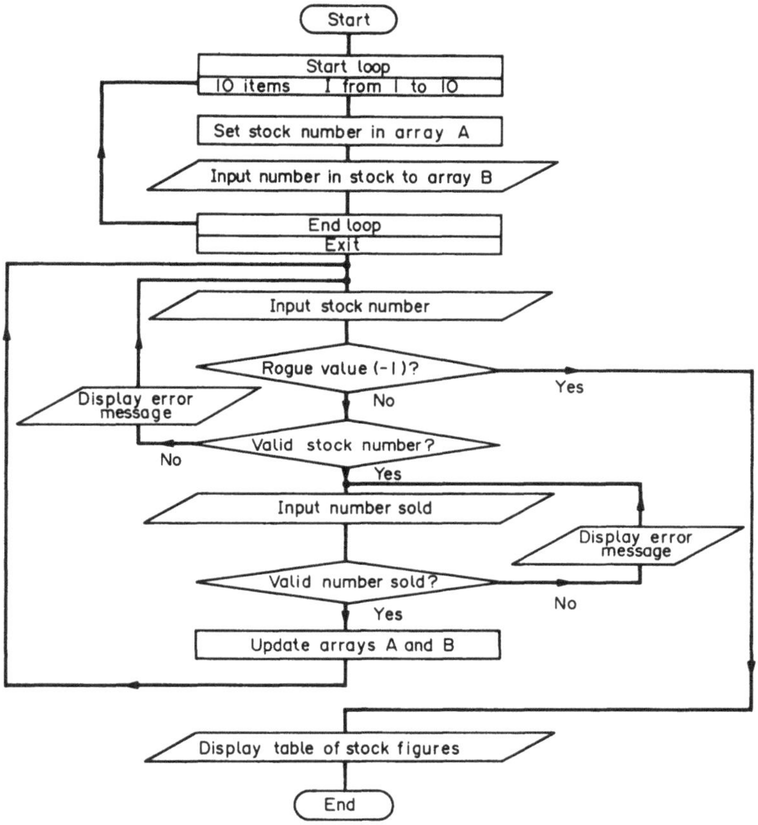

FIG. 10.1. Flow chart for BASIC program Example 1 (stock control).

```
?60
.....
etc. for 10 stock numbers
.....
PLEASE ENTER STOCK NUMBER FOR NEWSPAPER SOLD
?2
PLEASE ENTER NUMBER SOLD
?3
PLEASE ENTER STOCK NUMBER FOR NEWSPAPER SOLD
?-1
END OF DAY STOCK FIGURES
STOCK NUMBER..NUMBER IN STOCK
      1         40
      2         57
      3         30
      :          :
     10         60
```

BASIC Program Example 2

PROGRAM DESCRIPTION (FOR COMPETITION MARKS CALCULATION)

A gymnastics competition has five judges and thirty competitors. Each competitor's highest and lowest marks are discarded, and the three remaining marks are added to produce each competitor's total mark. A BASIC program must:

(a) input a competitor's name and the five marks for that competitor, for each competitor in turn
(b) display each competitor's name and total mark
(c) display the overall winner's name.

PROGRAM LISTING

```
10   REM GYMNASTICS COMPETITION RESULTS PROGRAM
20   REM INPUT 5 MARKS FOR EACH OF 30 COMPETITORS
30   LET TOPMARK=0
40   FOR I=1 TO 30
50   LET LOWEST=10;LET HIGHEST=0;LET TOTAL=0
60   PRINT "PLEASE ENTER COMPETITOR NAME"
70   INPUT N$
80   PRINT "PLEASE ENTER 5 MARKS"
90   REM APPLY A LOOP TO PROCESS 5 MARKS
100  FOR J=1 TO 5
110  INPUT MARK
120  LET TOTAL=TOTAL+MARK
130  IF MARK>LOWEST THEN GOTO 150
140  LET LOWEST=MARK
150  IF MARK<HIGHEST THEN GOTO 170
160  LET HIGHEST=MARK
170  NEXT J
180  REM NOW DISPLAY COMPETITOR'S NAME AND TOTAL
190  LET TOTAL=TOTAL-(HIGHEST+LOWEST)
200  PRINT N$," GAINED ",TOTAL," MARKS"
210  REM CHECK IF THIS IS THE BEST COMPETITOR SO
     FAR
220  IF TOTAL<TOPMARK THEN GOTO 250
230  LET TOPMARK=TOTAL
240  LET W$=N$
250  NEXT I
260  REM DISPLAY THE WINNER'S NAME
270  PRINT "WINNER IS ",W$
280  END
```

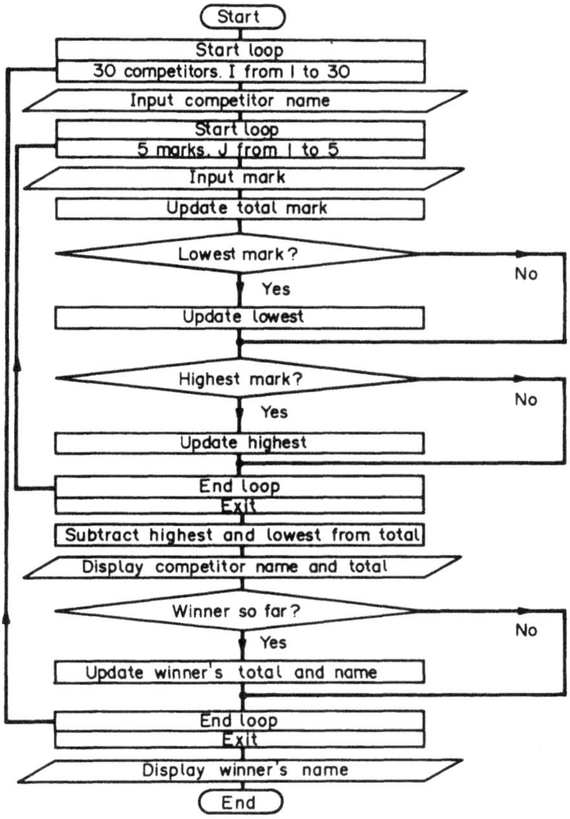

FIG. 10.2. Flow chart for BASIC program Example 2 (competition results).

FLOW CHART

See Fig. 10.2.

TEST DATA

Typical test data, together with the information displayed by the program, is:

```
PLEASE ENTER COMPETITOR NAME
L WILLIAMS
PLEASE ENTER 5 MARKS
?8
?6
?7
?3
```

```
?6
L WILLIAMS GAINED    19 MARKS
....
etc. for 30 competitors
WINNER IS    S THOMAS
```

10.3 Pascal

Pascal is a programming language that is fundamentally different to BASIC in its approach to solving a problem. Pascal forces the programmer to design his program in a far more orderly and modular manner. A BASIC program consists of a long list of sequential statements that are obeyed to perform the overall program function, whilst a Pascal program consists of an orderly block structure. In this way Pascal encourages "top-down" design and modularity, i.e. specifying the overall function and then breaking this down into separate tasks. An alternative description of Pascal is that it is a "structured" programming language, i.e. the program blocks or modules are chosen to be autonomous and easily understood and tested.

Most Pascal implementations use a straightforward compiler. A variation is the Pascal two-stage conversion process, that uses an intermediate "P-code", as follows:

```
Pascal source program → P-code → Machine code
                        ↑         ↑
     Conversion process   Conversion process uses a
     uses a compiler      "translator" (or an interpreter in the
                          same way as a BASIC interpreter)
```

The British Standard 6192:1982 specifies the syntax of the range of Pascal statements, but clearly variations between different Pascal implementations on different computers exist. In particular an increasingly popular variation of the language is "Turbo Pascal", which is now offered on a large number of personal computers. The program examples given here are written in Turbo.

Consider the following sample Pascal program:

```
PROGRAM Circlecircum;
(*A Pascal program to calculate the circumference
of a circle*)
CONST
     pi=3.14159
VAR
     radius, circum:REAL
BEGIN (*Main body of program*)
     WRITE('Please enter circumference in cm');
```

```
    READ(radius);
    circum:=2*pi*radius;
    WRITELN('Circumference is ',circum,' cm')
END.
```

Notice the following points:

(a) Pascal treats upper-case (capital letters) and lower-case characters in exactly the same manner, and so they are interchangeable. Case only normally matters when text is displayed on the monitor. In some implementations only upper-case can be used throughout all statements.
(b) Comments are inserted in parentheses (with asterisks).
(c) The use of additional spaces (and therefore indents) and new lines within Pascal statements is disregarded by the Pascal compiler. This allows the programmer to position parts of his statements in a form to make the program more readable. Statements are terminated by a semi-colon (;).
(d) All constants and variables used in the program must be specified before the main body of the program so that the Pascal compiler can reserve memory space for them. In this example a constant for "pi" is specified with the CONST declaration, and the variables "radius" and "circum" are specified with the VAR declaration. Notice that the numerical mode of the variables must be defined—REAL (floating point) in this case. Integer (whole) numbers can also be used.
(e) The main body of the program lies between the BEGIN and END statements. The four statements are each separated by a semi-colon. There is no requirement for the semi-colon after the last statement (the END commands acts as a delimiter), although a semi-colon at this point is optional in some versions of Pascal. The suffix "LN" after the WRITE (or READ) command denotes a new line. The sign ":=" means "takes the value of" and is equivalent to "=" in BASIC. Therefore the program displays a request message, reads in a value which the operator enters for the circle radius and displays the circle circumference.

The principal features of program modularity and structured approach of Pascal programs are illustrated in the following examples.

Pascal Program Example 1

Simple statistical values are often required to summarise long lists of numbers. The following program calculates and displays the average and standard deviation for a list of 10 numbers that are entered by the operator.

```
PROGRAM Statistics;
(*Calculate mean and standard deviation for 10
numbers entered*)
VAR
   count:INTEGER
   number,total,squares,average:REAL
PROCEDURE Getnumbers;
(*This procedure reads in 10 numbers*)
BEGIN
   FOR count:=1 TO 10 DO
      BEGIN
         READ(number);
         total:=total+number;
         squares:=squares+SQR(number);
      END;
END;
PROCEDURE Results;
(*This procedure displays the results*)
BEGIN
   average:=total/10;
   WRITE('Average is ',average);
   WRITELN('Standard deviation is
   ',SQRT(squares/10-SQR(average)));
END;
BEGIN (*This is the main program*)
   WRITELN('Please enter 10 numbers');
   Getnumbers;
   Results;
END.
```

Notice the following points:

(a) The main program is placed after the procedures. The procedures are simply modules of the overall program and consist of groups of statements. The procedure name, e.g. Getnumbers, forms a statement in the main program when that module is to be called. This program construction helps to make the program more readable. A variation of a procedure is a "function", which returns a value to the main program.
(b) Program loops can be created using the FOR command, as applied in the procedure Getnumbers. There are alternative ways to this FOR/TO/DO command structure for creating loops in Pascal, viz. REPEAT/UNTIL and WHILE/DO, but the FOR command is normally adequate for most programs that require repeated sections.

(c) Mathematical functions, e.g. SQR (square) and SQRT (square root), are available, as in BASIC.
(d) Reserved words are highlighted by spelling them in capital letters.

Pascal Program Example 2

A common requirement of a computer program is to sort a list of numbers into order of magnitude, e.g. to sort customers' names in order of their account numbers, to sort stock items into order of their numbers left in stock. The following program sorts 100 integer numbers, which are generated randomly, into order of ascending magnitude. The technique applied is the "bubble" sort, which consists of comparing adjacent numbers in the list and reversing each pair if it is not in the required order.

```
PROGRAM Sort;
(*Sorts 100 random numbers into ascending order*)
CONST
   Lastitem=100;
VAR
   list:ARRAY(1..100) OF INTEGER;
   item,temp,m,n:INTEGER
PROCEDURE Create;
(*This procedure creates 100 random numbers in an
array*)
BEGIN
   RANDOMIZE;
   FOR m:=1 TO lastitem DO
      BEGIN
         list(m):=RANDOM(500);
      END;
END;
PROCEDURE Switch;
(*This procedure reverses two adjacent items in the
array*)
BEGIN
   temp:=list(n);
   list(n):=list(n+1);
   list(n+1):=temp;
END;
PROCEDURE Output;
(*This procedure displays 100 items in the array
"list"*)
```

```
BEGIN
   FOR m:=1 TO lastitem DO
      BEGIN
         WRITE(list(m));
         WRITE('    ');
      END;
   WRITELN;
END;
PROCEDURE Sort;
(*This procedure sorts 1ØØ numbers in ascending
order*)
BEGIN
   FOR m:=2 TO lastitem DO
      BEGIN
         n:=m-1;
         WHILE n>Ø DO
            BEGIN
               IF list(n)>list(n+1) THEN
                  BEGIN
                     Switch
                     n:=n-1;
                  END
               ELSE n:=Ø;
            END; (*Finish WHILE statement*)
      END; (*Finish FOR/TO/DO statement*)
END;
(*Now for the main program*)
BEGIN
   WRITELN('Sort program');
   Create;
   Sort;
   WRITELN('These are the 1ØØ numbers in order');
 WRITELN;
   Output;
END.
```

Notice the following points:

(a) The array must be declared in the VAR statement.
(b) The IF/THEN/ELSE statement in the procedure Sort is a convenient way of selecting one of two different program functions.
(c) There are four procedures, one of which (Switch) is called from another (Sort), i.e. procedure "nesting" occurs.
(d) The reader may like to check the logic of the Sort procedure. The WHILE/DO loop sorts all elements in the array, beginning with 2

items, then 3 items, and so on. The outer FOR/TO/DO loop ensures that 100 items are sorted.

Pascal Program Example 3

Serious programming, unlike most BASIC applications, requires that data files on backing store are accessed for reading, writing or both. The following simple demonstration program illustrates the ease with which data files are accessed within Pascal programs. Several additional Pascal commands exist for file manipulation, and the reader is referred to the Bibliography for more detailed information.

```
PROGRAM Filedemo (INPUT,OUTPUT);
VAR  Count:INTEGER
   Total,Entry,Average,Stored:REAL;
   Testfile:FILE OF REAL
CONST  n=5;
BEGIN
   REWRITE(Testfile);   (*Opens a new file for
   writing*)
   Total:=0;
   FOR Count:=1 TO n DO
      BEGIN
         READ(Entry);   (*Input a number from
         terminal*)
         Total:=Total+Entry;
         WRITE(Testfile,Entry);   (*Writes Entry to
         file*)
      END;
   Average:=Total/n
   WRITELN('AVERAGE IS ',Average);   (*Displays
   average*)
(*We have now stored 5 manually entered numbers in
a file. Next read these numbers back off file*)
   RESET(Testfile);   (*Opens file for reading*)
   Total:=0;
   FOR Count:=1 TO n DO
      BEGIN
         READ(Testfile,Stored);   (*Reads value
         from file*)
         Total:=Total+Stored
      END;
   Average:=Total/n;
   WRITELN('AVERAGE FROM FILE IS ',Average);
   CLOSE(Testfile);   (*Stores file on disc*)
END.
```

Notice the following points:

(a) The optional INPUT,OUTPUT in parentheses on the first line specifies that normal input and output are via the terminal (rather than from a file).
(b) The file name must be specified in the VAR statement. Alternative file types to REAL are INTEGER and CHAR (text).

10.4 C

The C programming language has many similarities with Pascal, although it is not as "high" a high-level language as Pascal and other similar languages. It does require that the programmer has a more detailed knowledge of machine operation, but it produces extremely compact and efficient machine code.

C is a structured programming language, and it employs modularity (in the form of "functions"—program modules) in the same way as Pascal (in the form of "procedures" and "functions"). A simple C compiler is very small, whilst an external library of functions is required to perform input/output, file handling, etc.

C was written by Dennis Ritchie of Bell Laboratories around 1972, and it developed from its two forerunner languages BPCL and B, which in turn were based on Algol 60. It was written for UNIX-based PDP-11 minicomputers, but has followed the UNIX operating system (the master program in a multi-programming system) into applications with microcomputer systems. Indeed UNIX itself (described in chapter 13) and all system software required to support UNIX (compilers, editors, etc.) are now written in C. Although C is ideally suited for the preparation of such system programs, application programs (to solve engineering/commercial/scientific problems) are increasingly also being written in C instead of the so-called "problem-orientated" languages such as ALGOL and FORTRAN. This is particularly true of UNIX systems, which inherently have such a close relationship with C.

Three simple example programs are presented here. Clearly the reader must refer to the Bibliography or to the documentation for the particular implementation, which will probably be a UNIX system, for a more detailed tutorial for the generation of meaningful application programs. The first example uses only one "function" module, and this is the "main" program itself.

Program Example 1

PROGRAM DESCRIPTION

This program converts an operator-entered value for a temperature reading in Fahrenheit into Celsius, and displays the result.

```
/*Program to convert a Fahrenheit temperature
                                to Celsius*/     (1)
#include "stdio.h"                               (2)
main()                                           (3)
{                                                (4)
   float fahren,celsius;                         (5)
   scanf("%f",&fahren);                          (6)
   celsius=(5.0/9.0)*(fahren-32.0);              (7)
   printf("%4.0f %6.1f\n",fahren,celsius);       (8)
}                                                (9)
```

Typically this program (called "conv.c") is entered into a UNIX system using an editor (described in Chapter 13), and is compiled as follows:

```
%cc conv.c    ("%" is the UNIX prompt)
```

The resulting machine code program is named "a.out" and is executed as follows:

```
%a.out
```

The program waits for the operator to enter a Fahrenheit value (e.g. 70), and this displays:

```
70    22.2
```

The following notes relate to the line numbers shown alongside the program above—they are for explanation purposes only and are not part of the program:

(1) This is a comment, which is ignored by the compiler. Several lines of comments could be inserted between the /* and */ delimiters.
(2) This statement must be included because it refers to the standard library, and the program uses a standard library function (scanf).
(3) There is only one function (C program module) in this program, and this must be specified with the name "main". The parentheses following the function name enclose any arguments being passed to a function—there are none in this case.
(4) Together with (9) the brace symbols enclose the statements that comprise the function (like BEGIN and END in Pascal).
(5) All variables should be declared at the start of a function (or the start of the "main" function), e.g. the two floating-point variables "fahren" and "celsius" are to be applied in this program. Alternative data types to "float" are "int" (integer), "char" (character), and others.
(6) The "scanf" external library function inputs a number entered by the operator. The second argument within the brackets is the name of the variable "fahren", whilst the "&" symbol is obligatory in this case. The first argument which is enclosed between inverted commas " "

and beginning with "%" defines the data type for this variable—"f" for floating point in this case.

(7) This statement performs the numerical calculation. Notice that the assignment symbol is "=", not ":=" as in Pascal.

(8) The character output command "printf" formats and prints (displays on the console) character strings and numeric values, using the same argument structure as scanf. Notice that the first argument (before the first comma) defines the print mode for the two following arguments, i.e. fahren is to be printed as a floating-point number in a space of 4 characters with no characters after the decimal point, and celsius is to be printed in a space of 6 characters with 1 character after the decimal point. The "\n" characters denote a newline.

The following program example employs alternative techniques in C for performing similar tasks, and also extends the program to illustrate looping procedures.

Program Example 2

```
/*
** This program enters 10 values and converts them
from Fahrenheit to Celsius.
The results are displayed in a table.
**
*/
#include "stdio.h"
float fahren celsius;
main()
{
    int i;
    limit=500;                /*Set a constant*/
    printf("FAHRENHEIT CELSIUS\n");
    for(i=0;i<10;i++)
        {
        scanf("%f",&fahren);
        if(fahren>=limit)
            printf("Number is too large\n");
        else
        {
        calc;
        printf("%4.0f %6.1f  n",fahren,celsius);
        }
    }
}
```

```
/*Finished main function. Now for the other
function.*/
calc()
{
    #define CONV 0.55555    /*Conversion factor of
      5/9*/
    float temp;
    temp=fahren-32;
    celsius=CONV*temp;
}
```

In this example the program has been split into two functions—main and calc (positioned in the reverse order to that used in Pascal, when the main program appears last). Although the use of more than one function is a little laborious for this simple example, it does illustrate the program modularity that is used with C. The main differences between this program and Program Example 1 are:

(1) The variables fahren and celsius have their data types specified before the function "main" because they are used in both of the functions main and calc.
(2) The main function employs a "for" loop, which is processed 10 times. The first two arguments inside brackets within the "for" statement specify that the integer variable "i" is to be set to 0 on the first pass of the loop, whilst looping terminates when i reaches 9. The "i++" argument indicates that i is to be incremented by 1 on each pass of the loop. The 8 lines that constitute the "for" loop are contained within brace symbols. Each separate statement is terminated by a semi-colon.
(3) A multiple statement (if . . . else) is contained within the loop. This statement illustrates conditional execution—one of two paths is selected depending on the result of the check on the value of fahren (less than 500, or equal to/greater than 500). Notice that the second path (after "else") consists of two statements, which must be contained within braces.
(4) The function "calc" is called from "main". Variables, e.g. "temp", used within the function must be declared at the start of the function. Notice that the "define" statement sets up a constant value, and by convention capital letters are applied in the name of the constant ("CONV") in order to highlight its usage in the program. Normally parameters are passed between functions—their names are contained as arguments within the brackets following the function name.

Program Example 3

The previous two program examples were trivial examples of "applications programs". As explained earlier, C is more commonly used for the preparation of "systems programs". This program example illustrates in a simplified form the entry part of a word processor (or editor) program. As the operator enters each character of his text file, the program counts the numbers of characters, words and lines. Additionally the operator can trigger a test message display.

```
/*Simplified word processor program—character entry
section*/
main()
{
    int inchar,nuchars,nuwords,nulines;
    #define EOF -1;
                    /*Define End Of File character*/
    nuchars=nuwords=nulines=0;
    while((inchar=getchar())  !=EOF)
        {
        if(nuchars==0 && inchar='\n')
            test;
        else{
            ++nuchars;
            if(inchar==' '  || c=='\t'  || c=='\n')
                ++nuwords;
            if(inchar=='\n')
                ++nulines;
            }
        }
    printf("%d %d %d\n",nuchars,nuwords,nulines);
}
/*Function that displays a test message(A to Z,a
to z*/
test()
{
    char outchar;
    for(outchar='A';outchar<='Z';++outchar)
        putchar(outchar);
    printf('\n');
    for(outchar='a';outchar<='z';++outchar)
        putchar(outchar);
    printf('\n');
}
```

Notice the following features:

(1) A multiple statement at the beginning of the program sets three variables (nuchars, nuwords and nulines) to zero—these variables count the numbers of characters, words and lines.
(2) The "while" state includes an expression:

```
((inchar=getchar())  !=EOF)
```

which includes the use of a library function "getchar" to read an input character—its ASCII value is assigned to the variable "inchar". This "while" loop continues until the character entered is the End of File (EOF) character.
(3) The first "if" statement in the "while" loop illustrates the AND symbol "&&", such that the condition being checked is to determine if number of characters equals 0 *and* input character is newline. This implies that if the first character which the operator enters is newline, then the function "test" is obeyed.
(4) The second "if" statement in the "while" loop illustrates the OR symbol " || ", such that the condition being checked is to determine if the input character is a space, or a tab or a newline. If this condition is satisfied then the number of words is incremented.
(5) The "test" function displays all upper-case characters A to Z on one line, followed by all lower case characters a to z on a new line.

Note The program examples presented here can only give the reader a flavour of each programming language discussed. If the reader is to attempt other than straightforward variations of these programs, he/she must support this text with the reading list given in the following Bibliography, or by reference to programming manuals for his/her particular implementation.

Bibliography

1. *Interactive Computing with BASIC—a First Course.* Donald Munro, 1981.
2. *Computer Programming Languages in Practice.* C. A. Hofeditz. Heinemann, 1985.
3. *Basic Programming for Computer Literacy.* Donald Moursund. McGraw-Hill, 1978.
4. *Mastering Pascal Programming.* Eric Higgins. Macmillan, 1986.
5. *Introduction to Turbo Pascal.* Douglas S. Stivison. Sybex, 1985.
6. *Elementary Pascal.* Henry Ledgard and Andrew Singer. Fontana/Collins, 1982.
7. *The C Programming Language.* Brian W. Kerningham and Dennis M. Ritchie. Prentice-Hall, 1978.
8. *Introducing C.* Boris Allan. Collins, 1986.
9. *The C Primer.* Les Hancock and Morris Krieger. McGraw-Hill, 1982.
10. *A Book on C.* R. E. Berry and B. A. E. Meekings. Macmillan, 1984.

CHAPTER 11

THE CP/M OPERATING SYSTEM

11.1 Introduction

An "operating system" can be defined as the interface between the computer and the operator. It is the master program (or suite of programs) that controls the hardware resources of the computer and enables different application programs to be run by the operator. Effectively the operator gives commands to the operating system to enable programs (or "files") to be run, examined or altered. Before a disc-based computer can be used, its operating system must be running so that commands can be entered and obeyed.

CP/M (control program for microprocessors) was developed by Gary Kildall of Digital Research to support 8-bit floppy disc-based personal computers in the mid-1970s. It has now become the de facto standard for such machines, and although primarily written for 8080/8085/Z80 systems, CP/M is now available on machines that employ different CPUs. Several versions exist, but CP/M 2.2 is the most popular.

CP/M is a single-user operating system. A multi-user version of CP/M (called MP/M) has been introduced, and is described in section 11.8.

In a CP/M configuration there are effectively three program levels:

(a) the CP/M nucleus itself (the CP/M program processes operator commands, handles the hardware peripherals and performs file transfer to/from the backing store, which is normally floppy disc)
(b) utility programs, e.g. an editor to create new program files, assemblers and compilers—these are additional to CP/M
(c) applications programs, e.g. word processor, general ledger, payroll.

11.2 CP/M "Built-in" Commands

The normal hardware configuration of a CP/M system is shown in Fig. 11.1. When the machine is switched on (or when a RESET/RESTART pushbutton is pressed) a ROM-based "bootstrap loader" program transfers CP/M down from disc into memory and then enters CP/M. CP/M displays an operator prompt ("A>" if disc drive A is operational). The operator then enters commands via the VDU to activate a variety of programs, and the

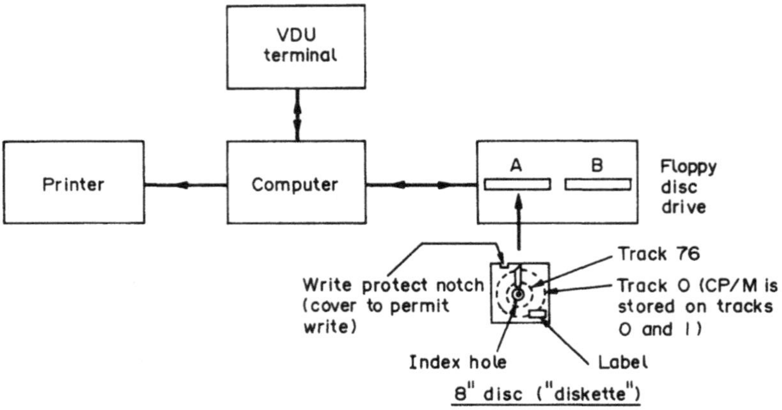

FIG. 11.1. Typical CP/M hardware configuration.

results can be indicated either on the VDU screen or on the printer. The 8-inch disc is often called a "diskette".

There are five principal "built-in" commands (sometimes called "resident" commands) within CP/M:

(a) DIR

This produces a directory of filenames, e.g.

```
A>DIR↙
```

Note ↙ denotes RETURN key. Underlined characters indicate operator entry.

produces

```
A:ED      COM:PIP      COM:ASM      COM:LOAD      COM
etc.
```

This indicates that the program files ED.COM, PIP.COM, ASM.COM and LOAD.COM are held in disc drive A. The .COM "extension" indicates the nature of the file—COM indicates a "command", or machine code, file. Other common file type extensions are .ASM (assembly language source listing) and .BAS (BASIC source listing). It is sometimes useful to use the "wildcard" symbol "*", as follows:

```
A>DIR *.COM↙
```

and CP/M responds by displaying all filenames ending with the .COM extension. Notice that:

```
A>DIR B:↙
```

lists all filenames on disc drive B.

(b) TYPE

This produces a listing of a file on the VDU screen or on the printer, e.g.

```
A>TYPE MAIL.ASM↙
```

produces a listing of the assembly language program MAIL on the VDU. In order to obtain the listing on the printer also, the operator must firstly press:

```
CONTROL P↙                (CONTROL is a single key)
```

When the printout is complete CONTROL P must be re-entered to turn off the printer for all subsequent listings and CP/M messages.

(c) ERA

This erases a file from the backing store, e.g.

```
A>ERA SORT.COM↙
```

erases the file SORT.COM. The use of wildcards is permissible with this command, e.g.

```
A>ERA FIND.*↙
```

erases all versions of the program FIND, e.g. FIND.ASM, FIND.COM and any others that may exist. The following command therefore erases all files on disc (except CP/M itelf):

```
A>ERA *.*↙
```

(d) REN

This renames a file, e.g.

```
A>REN HARRY.ASM=GEORGE.ASM↙
```

renames the file GEORGE.ASM as HARRY.ASM.

(e) SAVE

This stores an area of memory to disc, e.g.

```
A>SAVE 3 TOM.COM↙
```

saves the first three 256-byte pages from the start of the user memory to disc, and names the new file TOM.COM.

The operator can transfer disc control to drive B, as follows:

```
A>B:↙
```

and the prompt "B>" is then given by CP/M.

Figure 11.2 illustrates a typical memory map for a CP/M system. 48K or 64K of memory is required, and CP/M, which implements the built-in

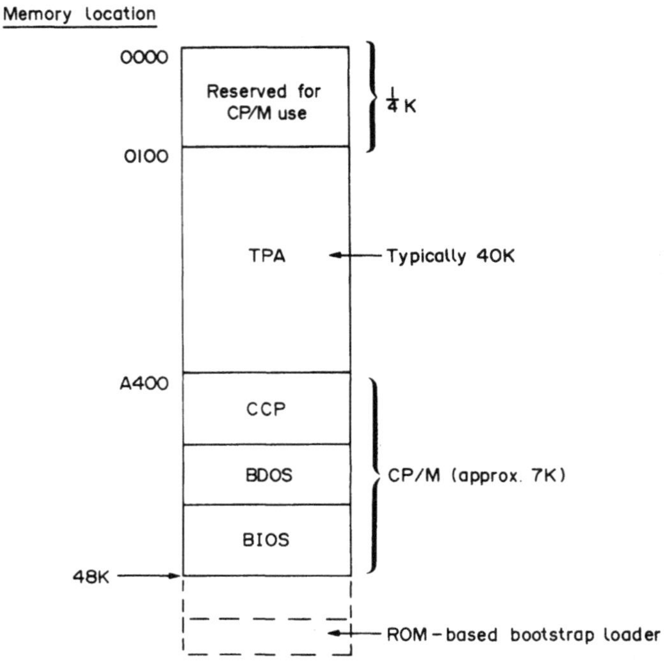

FIG. 11.2. Typical memory map for 48K CP/M system.

commands, resides at the bottom (high addresses) of memory. Notice that CP/M consists of three program modules:

CCP (console command processor)	interprets operator commands
BIOS (basic input/output system)	performs all input/output operations
BDOS (basic disc operating system)	handles all disc file transfers and allocation of disc space.

The unused area of memory in which CP/M runs all other programs is called the TPA (transient program area). It extends from memory location 256 (hex. 0100) to the start of CCP, and the TPA size depends on the amount of RAM available in the computer. Note that a user program is allowed to overwrite CCP, and when it is finished CCP must be reloaded from disc. This is termed a "warm boot" to distinguish it from a "cold boot", when the whole of CP/M is transferred from disc on start-up or reset/restart.

11.3 CP/M "Transient" Commands

The difference between transient commands and built-in commands is that the former call additional utility programs that must be held on disc. The principal transient commands are as follows:

(a) FORMAT

This prepares a blank disc for data storage, e.g.

```
A>FORMAT↙
```

transfers the program FORMAT.COM from disc drive A, and then enters it. FORMAT asks the operator several questions, including which drive contains the blank disc, whether single-sided or double-sided recording is to be applied and whether single-density or double-density recording is required. FORMAT then prepares all sectors (typically one sector is 128 or 256 bytes) around the disc surface, including the writing of each sector self-address (track and sector number) at the start of each sector. Note that formatting a disc removes any previous information on it.

(b) SYSGEN

This copies CP/M from one disc to another, e.g.

```
A>SYSGEN↙
```

copies tracks 0 and 1 from drive A to drive B.

(c) MOVCPM

This facility is not frequently used, but it allows CP/M to be adjusted for a different amount of memory within a different hardware configuration, e.g.

```
A>MOVCPM 48 *↙
```

alters CP/M (perhaps from a 32K system) to use 48K of memory. The * indicates that the new CP/M should be retained in memory, before being transferred to disc using SYSGEN.

(d) PIP

The peripheral interchange program (PIP) is used to copy files from one disc to another, e.g.

```
A>PIP B:PROVED.COM=A:TRIAL.COM↙
```

copies the file TRIAL.COM on disc drive A to disc drive B, giving the new file on B a name PROVED.COM. TRIAL.COM is retained on A. An entire disc (excepting CP/M, which must be copied using SYSGEN) can be copied using wildcards for filenames, as follows:

```
A>PIP B:=A:*.*↙
```

(e) STAT

This utility program gives information about the size of files on disc, e.g.

```
A>STAT↙
```

produces a statement indicating the amount of storage space available on the current drive, whilst:

```
A>STAT STOCKS.COM↙
```

returns information about the size of the file STOCKS.COM in both bytes and records (128 bytes). Notice that file sizes are in increments of 1K for single-density discs, and 2K for double-density discs.

(f) ED

The editor program ED allows text files to be created and altered, e.g.

```
A>ED CONTROL.ASM↙
```

allows the operator to create/examine/amend the assembly language file CONTROL.ASM. Editor facilities are described in detail in section 11.4.

(g) ASM

The 8080 assembler converts an assembly language file into a machine code file (in "hexadecimal" form), e.g.

```
A>ASM CONTROL.ASM↙
```

examines the assembly language file CONTROL.ASM and creates a file CONTROL.HEX (if no assembly errors occur). The Z80 assembler program ZASM can be used in place of ASM.

(h) LOAD

This converts the hex. file produced by the assembler into a final executable machine code file, e.g.

```
A>LOAD CONTROL↙
```

converts CONTROL.HEX into CONTROL.COM. This latter file can be called and run by CP/M by simply entering its name as follows:

```
A>CONTROL↙
```

(i) DDT

The dynamic debugging tool allows the programmer to test and debug a machine code program, e.g.

```
A>DDT CONTROL.COM↙
```

allows part, or all, of the program CONTROL.COM to be executed and CPU registers, etc. to be examined.

The Z80 debugger is named ZSID (in place of DDT).

(j) Interpreters and Compilers

Several interpreters and compilers are offered to support CP/M systems. A BASIC interpreter is available, as follows:

```
A>MBASIC↙
```

to operate on (interpret) a text file created using the editor and possessing a filename extension of ".ASC". Further, a Pascal compiler is available as follows:

```
A>TURBO HOUSE.PAS↙
```

to compile a file named HOUSE.PAS which has been created using the editor.

(k) SUBMIT

It is often necessary for the operator to enter a series of CP/M commands from the selection above. Perhaps an author may wish to use the editor to insert the same few lines of text into several files. The time-consuming process of repeatedly typing in the same commands can be avoided if a "batch file" of these commands is created using the editor. Assuming that the operator names this file NAME.SUB the file of commands can be executed as follows:

```
A>SUBMIT NAME↙
```

If a parameter (e.g. a filename) must be inserted into the batch file when it is executed, then the operator can specify that parameter using the "$" character as follows. He creates the following batch file using the editor:

```
ASM  1$.ASM
LOAD 1$
1$
```

and he executes the batch file using:

```
A>SUBMIT NAME CONTROL↙
```

Therefore the file CONTROL.ASM will be assembled (using ASM), loaded (using LOAD) and the machine code version executed.

Note A summary of the main CP/M commands is presented in Table 11.1.

11.4 The CP/M Editor

The CP/M editor ED is used to create text files and to amend text files. Such files are normally source programs, either in high-level language or in

TABLE 11.1 *Commonly used CP/M Commands (a = argument)*

Command	Description
DIR	Display names of files on disc
TYPE a	Display contents of file
ERA a	Erase/delete file
REN a a	Rename file
SAVE a	Save an area of memory to disc
FORMAT	Format blank disc
SYSGEN	Copy CP/M from one disc to another
MOVCPM	Adjust CP/M for different memory space
PIP a a	Copy file (or entire disc)
STAT (a)	Check disc space used
ED a	Editor (to create/alter) file
ASM a	Assembler for 8080 (use ZASM for Z80)
LOAD a	Loader
DDT a	Debugger for 8080 (use ZSID for Z80)
SUBMIT a	Submit (run) batch file

assembly language. Figure 11.3 illustrates the use of ED, and summarises the main commands used. A typical order in which the editor commands are used in order to alter an existing text file, e.g. an assembly language program, is:

(a) Enter ED, from CP/M, using:

```
A>ED FILENAME.ASM↙
```

The editor now runs and it displays a "*" character prompt.

(b) Append previous file:

```
*#A↙
```

where "#" denotes that the entire file should be appended (transferred from disc into a memory buffer).

(c) Display the current file:

```
*BØP↙
```

where B moves the line pointer to the beginning of the file and ØP displays the first screen page (1P would display the second page).

(d) Move to the required line where corrections are to be made:

```
*B8L↙
```

moves from the beginning of the file forward 8 lines, i.e. to line 9.

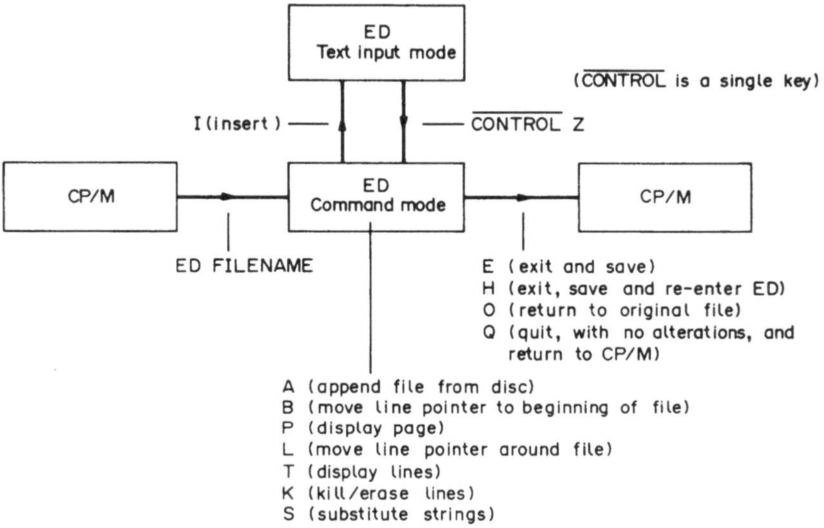

FIG. 11.3. Editor (ED) modes of operation.

(e) Kill/erase the current line and the line that follows:

*2K↙

removes lines 9 and 10.

Note Steps (b) to (e) are unnecessary when a new file is being created using a filename that has not been used before.

(f) Enter insert (text input) mode:

*I↙

and new lines of text, e.g. assembly language instructions, can be entered.

(g) Leave insert mode:

CONTROL Z

to re-enter normal editor command mode when the operator has finished inserting lines of text (each line is terminated by operation of the RETURN key).

(h) Store amended file:

*E↙

corrected file is stored on disc and control returns to CP/M. Variations of the E command exist—see Fig. 11.3. Notice that, when an updated file is stored back on disc in this manner, the previous version of the file is retained on disc and is given an extension of ".BAK". This back-up version could be useful if the amendments were incorrect and needed to be cancelled. The .BAK file could be renamed as the .ASM file to achieve this.

The editor commands listed above are the most commonly applied. However, others do exist to allow advanced editing functions to be performed.

A typical example of the use of the editor is as follows:

Example—to correct one line in an assembly language program file

```
A>ED LIGHT.ASM↙
*#A↙
*BØP↙
1.    ORG 1ØØH
2.    MVI A,1
3.    OUT (2ØH),A
4.    JMP Ø
5.    END
*B2L↙
*K↙
*I↙
   OUT (21H),A↙
CONTROL Z
*BØP↙
1.    ORG 1ØØH
2.    MVI A,1
3.    OUT (21H),A
4.    JMP Ø
5.    END
*E↙
A>
```

11.5 The CP/M Assembler

The assembler that is supplied with most CP/M systems is the 8080 assembler ASM.COM. The Z80 equivalent is named ZASM.COM. A typical 8080 program file that can be created using the editor, and then can be presented to ASM, is as follows:

```
         ORG 100H
INPORT   EQU 10H
OUTPORT  EQU 11H
         IN  INPORT    ;Input byte from input port to
accumulator
         ANI 1         ;Check bit 0
         JZ  NOTSET    ;Jump if zero
         LDA VALUE     ;Get output byte
         OUT OUTPORT   ;Output to output port
NOTSET:  JMP 0000H     ;Jump to CP/M
VALUE:   DB  40H
         END
```

The first three and last two lines ae not part of the actual program and are termed "pseudo-instructions". They are interpreted by the assembler as follows:

(a) ORG—"Origin" directive to set the absolute address of the program
(b) EQU—"Equate" directive to assign a numerical value to a label
(c) DB —"Data byte" directive to place data bytes into memory locations
(d) END—"End" directive instructs the assembler to stop at this line.

The Z80 assembler ZASM uses the same directives except for (c), which has the variations shown in Fig. 11.4. The asembler will report any errors, e.g. invalid mnemonic, label not specified in the source file. These errors must be corrected using the editor before the file is presented to the assembler again. If no errors exist, the assembler produces a "hex" file of machine code, which must be "loaded" using LOAD.COM to produce the final executable machine code file.

ASM (8080)	ZASM (Z80)	Meaning
DB 20	DEFB	Place data byte 20 (decimal) in memory location
DB 09H,0AH,0BH	DEFB	Place data bytes 09,0A,0B (hex.) in consecutive memory locations
DB 'ALARM'	DEFM	Place characters (ASCII) for A,L,A,R,M in consecutive memory locations
DW 2090H	DEFW	Place data word (16 bits) for 2090 (hex.) in consecutive memory locations
DS 32	DEFS	Reserve 32 (decimal) bytes in memory

FIG. 11.4. CP/M Assembler variations of DB directive.

11.6 The CP/M Debugger

Very few assembly language, and even fewer high-level language, programs work correctly first time. A debugger program allows the programmer to closely monitor the way a machine code version of a program runs, e.g. the program can be stopped at selected points, and CPU registers and memory locations can be examined.

The standard CP/M 8080 debugger program is named DDT (dynamic debugging tool), while a common Z80 debugger is termed ZSID. They possess almost precisely the same functions—the only significant difference is that DDT uses an operator prompt of "–" whilst ZSID uses "#". For this reason, only DDT is described here.

Assuming that DDT.COM is held on the system disc, the following CP/M command calls the debugger, requesting that the machine code program CLOCK.COM is to be tested:

```
A>DDT CLOCK.COM↙
```

DDT is transferred to the beginning of the TPA area, it then relocates itself upwards in memory such that location 0100 onwards is available for CLOCK.COM. The programmer can then test CLOCK using the following commands:

(a) List the program, e.g.

```
-L100↙
```

displays in disassembled form, i.e. using assembly language mnemonics, the machine code program commencing at memory location 0100 in blocks of approximately 12 lines/instructions. Repeated operation of the "L" key steps through consecutive blocks of the program.

(b) Display memory locations, e.g.

```
-D2000↙
```

displays in hexadecimal form a block of memory locations commencing at location 2000.

(c) Execute the program to a breakpoint, e.g.

```
-G100,116↙
```

runs the program commencing at location 0100, and stops the program at location 0116 (the instruction commencing at 0116 is not obeyed). Simply pressing the "G" key runs the program to completion.

(d) Examine the CPU registers, e.g.

`-X↙`

displays the state of all CPU registers (A, B, C, D, E, H, L, program counter, stack pointer, status flags).

(e) Trace several instructions, e.g.

`-T6↙`

obeys 6 instructions, performing an automatic "X" command after each, i.e. display CPU registers after each instruction is completed.

(f) Untrace several instructions, e.g.

`-U4↙`

obey 4 instructions, with no display function.

(g) Alter a CPU register, e.g.

`-XP↙`

displays the current contents of the program counter and waits for the operator to enter a new value, e.g.

`P=0116 104↙`

P can be replaced by different symbols for the other CPU registers.

(h) Alter a memory location, e.g.

`-S2005↙`

displays the current contents of location 2005 and waits for the operator to enter a new value, e.g.

`2005 29 20↙`

The debugger stays in this mode, stepping through consecutive memory locations, until period (".") is pressed.

(i) Assemble an instruction, e.g.

`-A109↙`

followed by:

`0109 MVI B.01↙`

in order to assemble the MVI instruction into machine code at location 0109. This mode is also terminated by a period.

11.7 Sample Procedure for Assembly Language Program

Step 1—Create file (program) using editor ED

```
A>ED SUM.ASM↙
NEW FILE
*I↙
     ORG 100H↙
     MVI B,5     ;Load B with 5↙
     MVI A,4     ;Load A with 4↙
     ADD B       ;Add B to A↙
     STO 0500H   ;Store answer at memory location
                  0500↙
     JMP 0       ;Return to CP/M↙
     END↙
CONTROL Z
*E↙
A>
```

Step 2—Assemble program, make corrections and re-assemble

```
A>ASM SUM.ASM↙
```

The assembler reports an error on line 5—the incorrect mnemonic of "STO" was used in place of "STA". This is corrected using the editor as follows:

```
A>ED SUM.ASM↙
*#A↙
*B4L↙
*K↙
*I↙
     STA 0500H   ;Store answer at memory location
                  0500↙
CONTROL Z
*BOP↙
     ORG 100H
     MVI B,5     ;Load B with 5
     MVI A,4     ;Load A with 4
     ADD B       ;Add B to A
     STA 0500H   ;Store answer at memory location
                  0500
```

```
        JMP 0           ;Return to CP/M
        END
    *E↙
    A>ASM SUM.ASM↙
```

Step 3—Load program

```
    A>LOAD SUM↙
```

Step 4—Debug program

```
    A>DDT SUM.COM↙
    -L100↙
    0100  MVI B,05
    0102  MVI A,04
    0104  ADD B
    0105  STA 0500
    0108  JMP 0000
    -G100, 108↙
    -X↙
    COZOMOEOIO A=09 B=0500 D=0000 H=0000
    S=0100 P=0108
    -S0500↙
    0500 09.↙
    -G0↙
    A>
```

11.8 MP/M

MP/M (Multiprogramming control Program for Microprocessors) is the multi-user version of CP/M. A typical hardware configuration is shown in Fig. 11.5. In this "network" only the master computer possesses a backing store (floppy disc in this case) and printer, and each slave computer/user shares these peripherals with other computers/users in the network. Clearly this arrangement is applied to save the cost of expensive peripherals. The disadvantage of the system is the time response when several users request the use of the printer or floppy disc simultaneously.

The master computer runs the MP/M program, which includes facilities for priority and time allocation for slave requests for peripheral handling. The slave computers run a version of CP/M, which does not possess BDOS but includes a communication program to handle data transfers to the master computer—this version is named CP/NOS. The master computer's

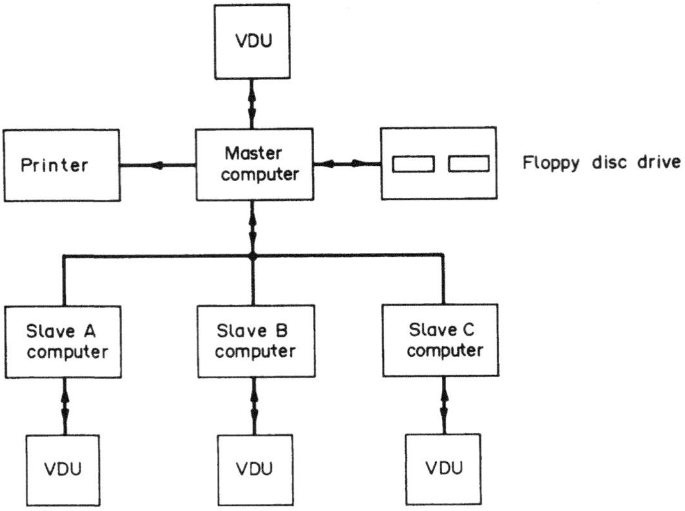

FIG. 11.5. MP/M hardware configuration.

MP/M offers the master user the following facilities which are additional to CP/M:

(a) Time-of-day display, which is set initially as follows:

```
OA>TOD Ø8/21/88 14:37:ØØ↙
```

This sets the time at 14.37 and the date at 21 August 1988. This is then updated by MP/M and can be displayed on a user's VDU.

(b) Program time scheduling, e.g.

```
OA>SCHED 1Ø/29/88 14:ØØ REPORT↙
```

schedules the program named REPORT to run at 14.00 hours on 29 October 1988.

Additionally MP/M gives the facility of message mailing, i.e. messages can be sent from one user's VDU to another.

Bibliography

1. *CP/M Primer.* Stephen Murtha and Mitchell Waite. Sams, 1983.
2. *Mastering CP/M.* Alan R. Miller. Sybex, 1983.
3. *CP/M—the Software Bus.* A. Clarke, J. M. Eaton and D. Powys Lybbe. Sigma, 1983.
4. *The CP/M Handbook with MP/M.* Rodney Zaks. Sybex, 1980.

CHAPTER 12

THE PC-DOS (MS-DOS) OPERATING SYSTEM

12.1 Introduction

PC-DOS, or to give it its other name MS-DOS, is to 16-bit desktop computers what CP/M is to 8-bit desktop computers. It is the operating system that has become an industry standard for single-user disc-based computers that employ 16-bit CPUs. The software company Microsoft spawned MS-DOS, as Digital Research had produced CP/M. MS-DOS was designed to be similar to CP/M, and this feature has been an important element in its acceptance to users. Although the two operating systems are not compatible, i.e. application programs written under CP/M will not run under MS-DOS, the similarity of the operator commands has enabled a CP/M user to adapt easily to MS/PC-DOS.

IBM, the world leader in the manufacture of mainframe computers (huge backing store, multi-multi-user, very fast CPU, wide range of peripherals), produced their first desktop computer system (the IBM PC) in the early 1980s. They chose the MS-DOS operating system and named it PC-DOS (personal computer disc operating system). These two operating systems are different in name only—in other respects they are virtually identical. The desktop computers that employ PC/MS-DOS are:

(a) IBM PC
(b) IBM PC-XT (as PC, but with hard disc replacing, or in addition to, floppy disc)
(c) IBM PC-AT
(d) Apricot
(e) Olivetti
(f) Compaq
(g) ACT Sirius
(h) DEC Rainbow

and several others.

The adoption of PC-DOS by the giant IBM is a guarantee of the widespread use of the system and the application program support for many years to come. Descriptions here will be biased towards PC-DOS applications due to the dominance of the IBM PC in the 16-bit market. For example, DOS version 1 is applied in (a) above, whilst DOS version 2 is applied on (b) to handle the larger storage capacity of the hard disc which allows several layers of file "directories". All versions of PC/MS-DOS are "upwards-compatible". Most of the first ranges of PC/MS-DOS systems were based on the Intel 8088 CPU (see chapter 6), but later models utilised other members of the Intel 16-bit family, e.g. the 80286.

The range of commercial, business and scientific software to support PC/MS-DOS is enormous, including such typical application packages as:

(1) word processor, e.g. Wordstar
(2) data base management, e.g. dBase 2
(3) spreadsheet, e.g. Supercalc, Lotus
(4) project planning

and so on.

12.2 PC/MS-DOS Nucleus

The analogy between CP/M and PC/MS-DOS is close in terms of program structure as well as operator commands. PC/MS-DOS can be broken down into three program modules, as follows for the IBM PC:

COMMAND (Command Processor)	—analogous to CP/M's CCP
BIO (Basic Input Output)	—analogous to CP/M's BIOS
DOS (Disc Operating System)	—analogous to CP/M's BDOS

Figure 12.1 shows the memory map for a typical PC/MS-DOS system. DOS and BIO are held in the lowest locations in memory, and COMMAND is divided around the user program area. The position of the transient part of COMMAND is dependent on the memory hardware configuration; its position adjusts to fill the end of usable memory (64K, 128K or even 256K). The system for bootstrapping the operating system consists of one extra process compared with CP/M. On start-up a ROM-based executive program transfers the bootstrap loader program from disc into memory. This program then runs to transfer PC/MS-DOS from disc into memory, and then enters the operating system. Normally the operating system boots off disc drive A, but even if the configuration includes a hard disc (containing the operating system) on drive C, the bootstrap loader will check drive A first. PC/MS-DOS gives the same operator prompt as CP/M ("A>" if drive A is the current system drive). System control can be transferred to a different disc drive, as follows:

```
A>C:↙
```

to transfer to drive C (typically the hard disc, if applied).

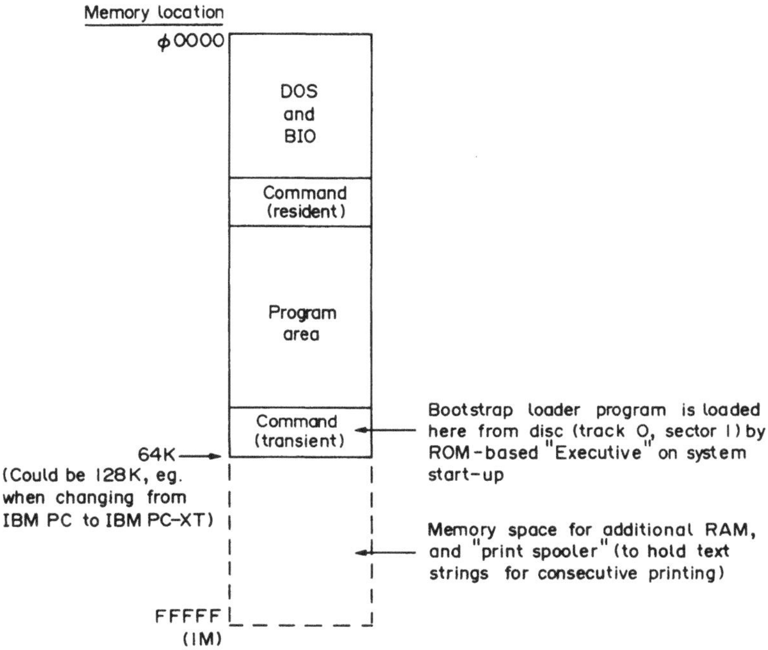

FIG. 12.1. Typical memory map for PC-DOS system.

Booting the system is achieved either by switching on the computer, or by pressing the CNTRL and ALT keys simultaneously and then momentarily pressing the DEL key. Some versions of PC/MS-DOS maintain an internal clock, and the operator may be asked to enter the date (month–day–year) and time (hours–minutes) before the A> prompt appears.

As with CP/M the nucleus of the PC/MS-DOS operating system contains several commands that allow files (principally programs) to be manipulated. The principal "internal" commands are:

(a) DIR

This produces a directory of filenames, e.g.

```
A>DIR↙
```

produces:

```
Volume in drive A has no label
Directory of A:\
FORMAT   COM    9Ø15  7-21-88
EDLIN    COM    7183  7-21-88
BASIC    COM   17Ø24  7-21-88
etc.
```

Programs	
.ASM	Assembly language program (in source code)
.BAK	Back-up file
.BAS	BASIC program (in source code)
.BAT	Batch command file
.COB	COBOL program (in source code)
.COM	Executable machine code program
.EXE	Relocatable program
.FOR	FORTRAN program (in source code)
.LIB	Library program
.OBJ	Machine code program (not in executable form)
.OVL	Overlay program
.PAS	Pascal program (in source code)
Work files (data or text)	
.DAT	Data file
.DOC	Document file
.TXT	Text file (same as .DOC)

FIG. 12.2. Filename extensions for PC/MS-DOS.

Wildcards are allowed, as with CP/M, e.g.

```
A>DIR *.COM↙
```

displays the names of all COM (executable machine code) files.

Figure 12.2 lists typical filename extensions, e.g. COM above, which describe the nature of files. An extension to a filename is not essential, but it does help to make the filename more meaningful.

(b) TYPE

This produces a listing of a file on the VDU screen or printer, e.g.

```
A>TYPE MEMO.TXT↙
```

produces a display of the text file MEMO on the VDU. The listing will appear on the printer also if the operator firstly presses:

`CNTRL P or CNTRL PRT.SC.` (for IBM PC)

An alternative command which produces a VDU listing only is:

```
A>COPY MEMO.TXT CON↙
```

or a printer listing only:

```
A>COPY MEMO.TXT PRN↙
```

Notice that with most of these file manipulation commands, the operator can specify which disc drive holds a particular file, e.g.

```
A>TYPE B:MEMO.TXT↙
```

indicates that the file is located on disc drive B.

(c) ERASE

This erases a file from a directory, e.g.

```
A>ERASE MONTH.COM↙
```

erases the file MONTH.COM from the current directory. An alternative command that performs precisely the same function is "DEL", e.g.

```
A>DEL MONTH.COM↙
```

(d) RENAME

This renames a file, e.g.

```
A>RENAME PREVIOUS.BAS PRESENT.BAS↙
```

changes the name of the file PREVIOUS.BAS to PRESENT.BAS. RENAME can be shortened to REN on some implementations.

(e) DATE

This displays the current date, e.g.

```
A>DATE↙
```

or it allows the date to be altered, e.g.

```
A>DATE mm-dd-yy↙
```

(f) TIME

This displays the current time, e.g.

```
A>TIME↙
```

or it allows the time to be altered, e.g.

```
A>TIME hh:mm:ss:ff↙
```

(g) CLS

This simply clears the VDU screen.

(h) VERIFY

This verifies after each disc write operation that data was transferred correctly, e.g.

```
A>VERIFY ON↙
```

sets the facility on, and

```
A>VERIFY OFF↙
```

sets the facility off.

(i) COPY

This command copies a file, either from one disc drive to another, or elsewhere on the same drive, e.g.

```
A>COPY CHARLES.COM BERT.COM↙
```

copies CHARLES.COM to another position on the current disc, and names the second file BERT.COM.

An example of a command that copies a file to another disc is:

```
A>COPY CHARLES.COM B:↙      (copies to drive B)
```

or

```
B>COPY A:CHARLES.COM↙       (copies from drive A to drive B)
```

Wildcards are allowed, so the following command copies all .COM files on drive B to A:

```
A>COPY B:*.*↙
```

Amongst the additional internal commands, there are commands that process directories—these are described in the next section. The "external" commands require supporting utility programs, and they are described in section 12.4

12.3 Directories and Sub-directories

When CP/M is booted, a single file-directory exists. Similarly a single directory is available when a new PC/MS-DOS system is used. However, the use of a hard disc ("Winchester") with several DOS implementations, e.g. the IBM PC-XT, allows very large numbers of files to be created and held on backing store. Implementation of the DIR command could produce an almost interminable and unmanageable list of filenames. For this reason PC/MS-DOS falls some way between CP/M and UNIX (the multi-user operating system described in the next chapter), and offers the facility of partitioning the file space into more than one directory, with some directories having sub-directories if required. This leads to a tree structure, and hence the master directory is sometimes called the "root" directory.

Figure 12.3 shows a typical PC/MS-DOS file structure. After booting the system, the root directory may look like:

```
A>DIR↙
Volume in drive A is USER1
Directory of A:\
FORMAT   COM    7028  2-18-88
DISKCOPY COM    4392  7-09-88
etc. (several PC/MS-DOS commands)
```

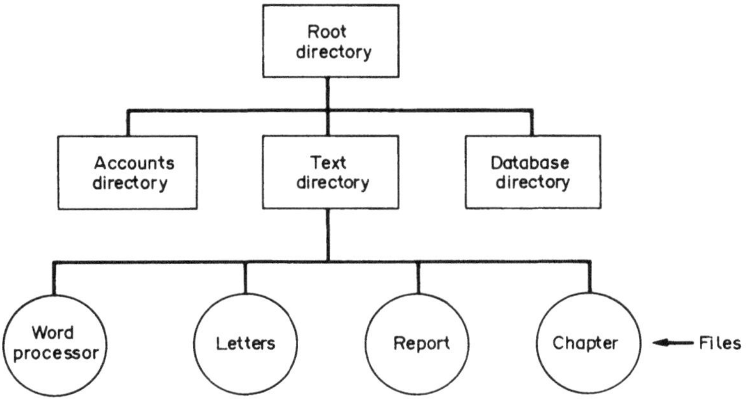

FIG. 12.3. Sample directory system.

```
ACCOUNTS <DIR>       12-20-87
TEXT     <DIR>       12-20-87
DATABASE <DIR>       12-22-87
```

This indicates that several PC/MS-DOS command files exist, as well as three sub-directories. These sub-directories sensibly group files that form an autonomous application area for the system. The diagram shows that the word processing sub-directory can be examined, and typically may look like:

```
A>DIR \TEXT\↙
Volume in drive A is USER1
Directory of A:\TEXT\
WSTAR     COM  15291  8-13-88
LETTERS   TXT    803  8-24-88
REPORT    TXT   9436  8-24-88
CHAPTER   TXT  10902  8-29-88
```

It is possible to have even further file levels of course, e.g. if LETTERS was a directory rather than a file. The "pathname" of any file or directory lists in order the directories that must be traversed to locate that file, e.g.

```
TEXT\CHAPTER.TXT
```

is the pathname for one of the text files in Fig. 12.3.

The backslash character is used to separate directory names. Also the backslash character at the start of a pathname is used to change back to the root directory. Therefore the following command allows an operator working in the ACCOUNTS directory to operate on a file in the TEXT directory:

```
A>ERASE \TEXT\REPORT.TXT↙
```

There are four commands in PC/MS-DOS that allow directory manipulation, as follows:

(a) CHDIR (or CD)

The CHDIR (Change Directory), or the alternative form CD, changes the current directory, e.g.

```
A>CHDIR ACCOUNTS↙
```

changes control from the root directory to the ACCOUNTS directory. Variations are:

```
A>CHDIR \ACCOUNTS↙
```

changes from any other directory back to root (backslash) and then to ACCOUNTS, and

```
A>CHDIR \↙
```

returns from a sub-directory to root.

(b) MKDIR (or MD)

The MKDIR (make directory) command, or the alternative form MD, creates a sub-directory, e.g.

```
A>MKDIR TEXT↙
```

Files can now be copied into this new directory.

(c) RMDIR (or RD)

The RMDIR (remove directory), or its shorter alternative form RD, removes a sub-directory, e.g.

```
A>RMDIR TEXT↙
```

All files in this directory must be erased before this command can be implemented. It is not possible to remove the root directory.

(d) TREE

This command displays all the directories and sub-directories on a disc, e.g.

```
A>TREE B:↙
```

12.4 External Commands

The PC/MS-DOS external commands require supporting files (utility programs) on disc. The main external commands are:

(a) FORMAT

This prepares a blank disc for data storage, e.g.

```
A>FORMAT↙
```

formats (prepares all tracks and sectors) disc A. As this destroys the current disc (erases FORMAT.COM, etc.), it is usual to format a different drive, e.g.

```
A>FORMAT B:↙
```

It is normally essential that drive C (hard disc) is not formatted, and it is prudent therefore to remove the FORMAT program from most discs.

Normally it is required to copy PC/MS-DOS itself onto a new disc, and this is achieved during the formatting process by adding an option to the FORMAT command, as follows:

```
A>FORMAT B:/S↙
```

An alternative, or additional, option to S is V, which allows the operator to give a label to a floppy disc, i.e. each floppy disc has a different name.

(b) DISKCOPY

This command is an alternative to FORMAT, but additionally it copies the entire contents of one drive to another, e.g.

```
A>DISKCOPY A: B:↙
```

copies all files from drive A to drive B, and formats the disc if necessary.

If the utility program DISKCOPY.COM is unavailable, the operator can use FORMAT and COPY (processing one or more files) to achieve the same result.

(c) BACKUP

This command backs up (copies) the contents of the hard disc (invariably drive C) to floppy disc (normally drive A), e.g.

```
A>BACKUP C:\ A:/S↙
```

copies all files and directories from drive C to drive A. All floppy discs on A must be formatted, and several discs will normally be required (e.g. 1 floppy disc = 360K bytes, hard disc = 10M bytes).

(d) RESTORE

This command reverses the process of the BACKUP command, e.g.

```
A>RESTORE A: C:\S↙
```

transfers all files and directories from the backup discs in drive A to the hard disc C.

(e) DISKCOMP

This command compares the complete contents of two floppy discs, e.g.

```
A>DISKCOMP A: B:↙
```

(f) COMP

This compares a single file (unlike DISKCOMP which compares entire discs), e.g.

```
A>COMP LETTER1.TXT LETTER2.TXT↙
```

(g) CHKDSK

The CHKDSK (check disc) command is similar to CP/M's STAT, e.g.

```
A>CHKDSK B:↙
```

returns a display of the number of files on drive B and the number of bytes free.

(h) EDLIN

The line editor program EDLIN enables files (e.g. programs) to be generated or altered, e.g.

```
A>EDLIN COMPUTE.PAS↙
```

allows the operator to create or amend a text file COMPUTE.PAS, i.e. a Pascal source program.

(i) Batch file

A batch file consists of a sequence of PC/MS-DOS commands. The batch file can be created using the editor, and is given a name, which must terminate with ".BAT". When the name is entered, as if it is a standard PC/MS-DOS command, the contents of the file are obeyed. The use of a batch file can save the operator entering the same sequence of commands repetitively. Consider the example of a batch file that formats a blank disc and then copies a series of files, e.g. a word processor program (WSTAR.COM) and two documents/text files, to it, as follows:

```
FORMAT B:/S
COPY WSTAR.COM B:
COPY REPORTA.TXT B:
COPY REPORTB.TXT B:
DIR B:
```

This list of commands can be entered as a file using the EDLIN editor,

TABLE 12.1 *Comparison of CP/M and PC/MS-DOS Commands (a = argument)*

CP/M	PC/MS-DOS	Description
DIR	DIR	Display names of files on disc
TYPE a	TYPE a	Display (optional print) contents of file
ERA a	ERASE a	Erase/delete file (or DEL for PC/MS-DOS)
REN a a	RENAME a a	Rename file
SAVE a	—	Save an area of memory to disc
—	DATE	Display date
—	TIME	Display time
—	CLS	Clear display
—	VERIFY	Confirm correct write to disc
PIP a a	COPY a a	Copy file (or entire disc)
ED a	EDLIN a	Editor (to create/alter a file)
STAT	CHKDSK	Check disc space used
FORMAT	FORMAT	Format blank disc
—	DISKCOPY	Copy entire disc
—	BACKUP	Dump hard disc to floppy
—	RESTORE	Reinstate hard disc from floppy
—	DISKCOMP	Compare two discs
—	COMP a a	Compare two files
—	CD a	Change directory (or CHDIR)
—	MD a	Make directory (or MKDIR)
—	RD a	Remove directory (or RMDIR)

perhaps with a name of TRANSFER.BAT. When the filename TRANSFER is subsequently entered, the sequence of commands is obeyed. A new disc is created and its directory is displayed.

Parameters can be passed into a batch file when it is executed, in exactly the same way as applied with the SUBMIT command in CP/M.

If a batch file is given the special name of "AUTOEXEC.BAT" then it is automatically executed each time the system is booted.

Note A summary of the main PC/MS-DOS commands, together with their CP/M equivalents (where appropriate), is listed in Table 12.1.

12.5 Line Editor

Files can be created, or altered, using the line editor EDLIN. Files can be programs or text files. An alternative to an editor is a word processor program, e.g. Wordstar.

Operation of EDLIN is very similar to that of CP/M's editor ED. It is called as follows:

```
A>EDLIN FILENAME.PAS↙
```

The text file (a Pascal program named FILENAME in this case) can then be generated, or altered if it already exists. The line editor displays a "*" prompt, and examples of the main commands are:

*3,23L↙	lists the file between lines 3 and 23
*16,19,28M↙	moves lines 16 to 19, and inserts before line 28
*1,10,16C↙	copies lines 1 to 10, and inserts before line 16
*2,4D↙	deletes lines 2 to 4
*5I↙	inserts 5 lines
*E↙	end edit, saving amended file and return to PC/MS-DOS
*Q↙	quit edit, without saving file

Bibliography

1. *The PC DOS Companion*. Stephen M. Murtha and Guy R. Petrie. Sams, 1983.
2. *MS/PC-DOS Prompt.* Randall McMullan, 1986.
3. *MS-DOS and PC-DOS on the IBM PC.* Charles Jackson. Glentop, 1986.

CHAPTER 13

THE UNIX OPERATING SYSTEM

Note UNIX is offered on a wide range of microcomputer and mini-computer systems. The programs in this chapter were tested on the following microcomputer:

MCS 68000 CPU board (with memory management unit)
80M byte Winchester, plus 1M byte floppy disc (8″)
$1\frac{1}{2}$M bytes RAM (256K for UNIX, $1\frac{1}{4}$M for users)
4 serial drive cards, each with 8085 CPU and 4 channels
  16 serial channels supply:
    10 terminals
    1 master terminal
    1 printer
    1 plotter
    1 computer link (to BBC computer)
    2 spare

13.1 Introduction

UNIX is a disc-based operating system that has become an industry standard for multi-user 16-bit microprocessor systems. Just as CP/M has become the standard for 8-bit single-user microprocessor systems, and PC/MS-DOS has become the standard for 16-bit single-user systems, so has UNIX dominated the 16-bit multi-user office computer market. It was originally developed by Bell Laboratories (Ken Thompson and Dennis Ritchie were the main architects) at around 1970 for minicomputers. Whilst it was applied on DEC PDP 11 minicomputers (PDP 11/45 and PDP 11/70) throughout the 1970s, it has now achieved even further use on the DEC VAX so-called "midicomputers" (or "super-minis"). With this strong historical background it was perhaps natural that it would be re-written to support 16-bit microprocessor systems based on the 8086, Z8000 and MCS68000. It is written in the high-level language C, which means that it can be implemented simply on any machine that possesses a C compiler.

UNIX is the world's most popular operating system for multi-programming time-sharing systems (CP/M supports only uni-programming systems).

Many derivatives of it are marketed under licensing agreements with Bell Laboratories, e.g. XENIX, CROMIX, ZEUS, ISIS. UNIX is particularly popular in universities around the world for teaching computer science students a variety of programming languages and concepts. As well as handling the computer's hardware resources and providing job (program) scheduling, the UNIX operating system includes a wide range of utility programs which are normally separate to other operating systems. These in-built utilities include editors and compilers, e.g. for Pascal, C, FORTRAN 77, BASIC, COBOL.

The fundamental differences between the single-user operating systems (CP/M and PC/MS-DOS) and UNIX are that UNIX has the two following characteristics:

(a) "multi-user", i.e. many users via their terminals can request programs and use the computer's resources
(b) "multi-tasking", i.e. a user can perform several tasks simultaneously, e.g. entering/editing a file, printing out a file.

13.2 UNIX File Structure

Whilst CP/M possesses a single directory which lists all the names of the files/programs in the system, UNIX allows each user to possess a separate directory, and indeed a user can establish other directory layers (sub-directories) if he wishes. Figure 13.1 illustrates the file structure possible within UNIX. Each user has a home directory; normally the name of this directory is the login name (the name the user enters when he commences to use the system). Home directories can have sub-directories, and any directory can be a mixture of sub-directories and files. This diagram resembles an inverted tree (home directories are branches, sub-directories are smaller branches, files are leaves and the misnamed root is the trunk). Since there can be several levels of directories a file name is fully expressed as its "pathname", as follows:

(a) /usr/smith/accounts/tax
(b) /usr/smith/accounts/sales
(c) /usr/smith/accounts/current
(d) /usr/smith/chapter2
(e) /usr/thomas/letters
(f) /usr/thomas/current
(g) /usr/jones/ledger ← directory only (no file)
(h) /usr/jones/results

Notice that lower-case letters (non-capital letters) are used for UNIX commands and filenames.

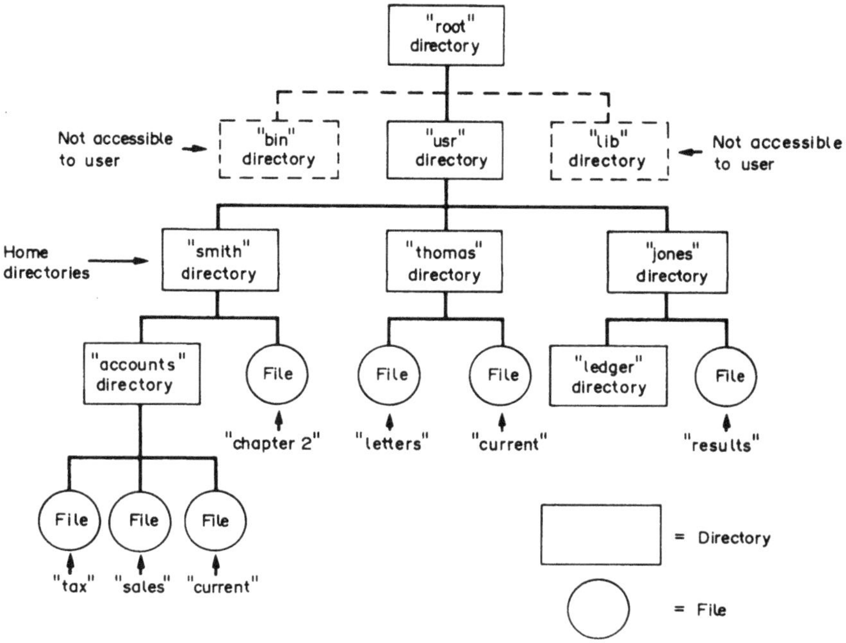

FIG. 13.1. UNIX file structure.

When the user is working in a particular directory he can access a file with a reduced filename, e.g.

/accounts/current

if the smith directory is currently accessed (the file/program can be executed but not listed in this case in order to provide some measure of security). Notice that the first slash (/) in the pathname stands for the root directory (if omitted the pathname is in the user's current directory), whilst all subsequent slashes serve to separate the name of a directory from the name of another directory or file.

Useful UNIX commands to assist the user in directory manipulation are:

(a) mkdir (make directory)
e.g. if jones is in his home directory and he enters:
mkdir reports
he succeeds in creating a new directory with a pathname
/jones/reports

(b) cd (change directory—called chdir in some early versions of UNIX)
e.g. cd /smith/accounts

causes a user in the jones directory to switch his current working directory to /smith/accounts.

Note that

cd . .

moves the user up one directory level.

(c) rmdir (remove directory)

e.g. rmdir ledger

enables a user working in his home directory jones to remove the ledger directory (only if there are no files in it—the next sections describe how to create and remove files).

(d) pwd (print working directory)

e.g. pwd (no argument is required) causes UNIX to display the full pathname of the current working directory. This is extremely useful when a user has lost track of exactly which directory he is currently working in. For example, if a user is currently working in the ledger directory, and he enters:

pwd

UNIX responds with:

/usr/jones/ledger

13.3 Handling Files

When UNIX is waiting for a command for an operator, it displays:

% (or $)

on the terminal. The user can then enter one of over 200 commands to the UNIX "shell". The shell is the operator command processor part of UNIX, and there are two versions of this shell program—the C Shell produces the % prompt, and the Bourne Shell produces the $ prompt.

An important feature of UNIX is that files consist only of the stream of byte-characters that the user enters. User files have no defined structure, and can be documents, computer programs or anything else that the user requires. Section 13.6 describes how to create a file using an editor program, but the following commands allow files to be examined and manipulated:

(a) ls (list contents of directory)

e.g. ls

lists on the operator's display all the names of the files and sub-entries in the current working directory.

ls ledger

uses an "argument" of ledger, and shows what is held in another directory—ledger in this case.

A useful variation of this command is:

ls −l

which gives a long listing, with directory entry being defined as follows:

```
total=30
drwxrwxrwx 1 smith  72 Feb 23 16.35 accounts
-rwxrwx--x 1 smith 648 Mar Ø2 Ø9.43 chapter2
```

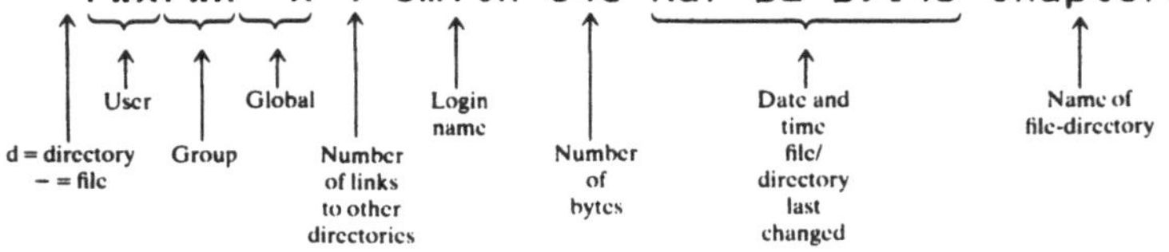

r = read
w = write
x = execute

} "permissions" to read and use files

(b) cat (read file)
e.g. cat letters
produces a listing of the file letters on the terminal. If the filename used is not present in the directory, use CONTROL D to stop UNIX attempting to create a file.

If the file is long, the cat command fills the screen and then adds more lines at the bottom, scrolling text off the top of the screen. A variation of cat exists to prevent this, as follows:

more letters

Press SPACE (the spacebar) to advance the text by one screenful, and press the RETURN key to advance by one line.

Note The ls and cat commands are probably the most commonly used UNIX commands.

(c) rm (remove file)
e.g. rm gloria
removes the filename gloria from the current working directory.

(d) cp (copy)
e.g. cp ordered purchased
copies the file ordered into a file purchased (ordered is retained).

(e) mv (move)
e.g. mv experiment results
changes the name of a file called experiments to results. The same command is used to change the names of directories.

(f) pr (print)
e.g. pr complaint

prints the file complaint on the system printer.
A variation of this command is:

lpr complaint

which "spools" the file complaint to the printer, such that the file is placed in a queue and printed when the printer is available.

A variety of other file handling commands exist, e.g. to merge files (sort), search a file for a particular pattern (grep), count the number of characters in a file (wc).

13.4 Further Facilities in UNIX Shell

Many UNIX shell commands require an input and/or output, and unless otherwise specified both are assumed to be the operator terminal, e.g.

cat analysis6

outputs the file analysis6 to the terminal.
However it is possible to redirect inputs and outputs as follows.

(a) > operator (redirect output to a file), e.g.

cat part1 part2> part3

This command concatenates (joins) file part2 to part1 and outputs the result to part3, which is a new file, rather than to the terminal.

(b) ≫ operator (redirect an append output to a file), e.g.

cat puppy≫dog

This command adds the contents of the puppy file to the dogs file.

(c) < operator (redirect input from a file), e.g.

ed<memo

This command redirects the input of the editor (used to create files—see section 13.6) to come from the disc file memo and not from the terminal.

(d) ¦ operator (pipeline), e.g.

ls¦wc –w (¦ is a vertical bar with a gap; on some keyboards it is |)

This command "pipes" the output of the ls command (the list of all the filenames in the current working directory) to the wc –w command (count the number of words). Only the output of the wc –w is seen on the operator terminal; the intermediate output of ls is not displayed. The output of the first command is the input to the second command, and this feature, together with the ability to combine redirection operators within a single command, makes UNIX extremely flexible.

Some other useful shell commands are:

date to display the current time and date
who to find out which other users are currently logged in to the system

learn to access a self-teach facility on the operation and use of UNIX
man to give information about each UNIX command ("on-line manual")

13.5 Logging in and Sending Mail

When UNIX is switched on and set running (by "bootstrapping" UNIX from disc to memory), the following message is displayed on each user terminal:

login:

The user must enter his login name before he can access UNIX facilities. UNIX will then ask for the "password", as follows:

password:

Whilst the user then enters his password, the letters he enters will not be echoed back to the display in order to prevent unauthorised personnel from acquiring that password and hence gaining access to the system. If the correct password is entered, UNIX replies with the prompt:

% (or $)

and the user can proceed to use the commands and facilities described in the previous sections. When the user has finished his work on the system, he must log out by entering the following:

logout

If the user does not have the message "login:" displayed when he comes to his terminal, he can press CONTROL D (i.e. press the D key while holding down the CONTROL key) to reinitialise UNIX for his terminal.

When logged in, a user has the useful facility within UNIX of sending and receiving mail between his terminal and other users. Immediately after login, UNIX will display to a user the following if messages have been sent to him by other users:

you have mail

The user can then read these messages by entering:

mail

UNIX replies giving details of the number of messages, and prompts with a "&" character. To read his mail the user enters:

p (to read the first message)
d (to delete the first message)

and so on for each message. The user quits mail with:

q

The user can also send mail to other users. Firstly he may like to check which users are currently using the system by entering:

who

He can then send a message to jones by entering:

mail jones

and no prompt or response from the system is given. He now enters his message, which can be several lines long, and terminates it by entering:

CONTROL D (CONTROL is a single key)

The user jones will be notified that mail is available for him when he logs in or requests mail. It is therefore possible to send a message to yourself, perhaps for the purpose of giving yourself a reminder when you next log in.

A more direct form of communication exists between users using the "write" command, such that a form of conversation is possible between users. If the following command is entered:

write jones

by the user whose login name is thomas, then the following is displayed on jones' terminal:

message from thomas

Thomas then enters his message and as he completes each line with operation of the RETURN key, that line appears on jones' terminal. CONTROL D is used to terminate a write sequence. Jones can reply if he wishes by commencing with:

write thomas

It is possible to block the mail and "write" facilities if a user wishes to work uninterrupted. The following commands block this electronic transfer of messages:

biff n (block "mail")
mesg n (block "write")

whilst the following commands permit message transfers

biff y (permit "mail")
mesg y (permit "write")

13.6 Screen Editor (vi)

UNIX possesses two editor programs which are used for creating files, e.g. high level language programs, as follows:

(1) ed—this is a line editor program, which allows the user to operate on only one line in that file at a time and thus can prove to be rather cumbersome when handling large text files

(2) vi —this is a screen editor program, such that it allows the user to move the cursor to any part of the text in order to work on that section.

The vi editor is a much more sophisticated and useful editor than ed, and for these reasons it is described here in preference to ed. It allows the user to examine the text file he is creating through a "window" (of typically 20 lines), which can be scrolled backwards and forwards through the file. The user can move the cursor anywhere within that window in order to make changes.

Figure 13.2 illustrates the modes of operation of vi, and also summarises

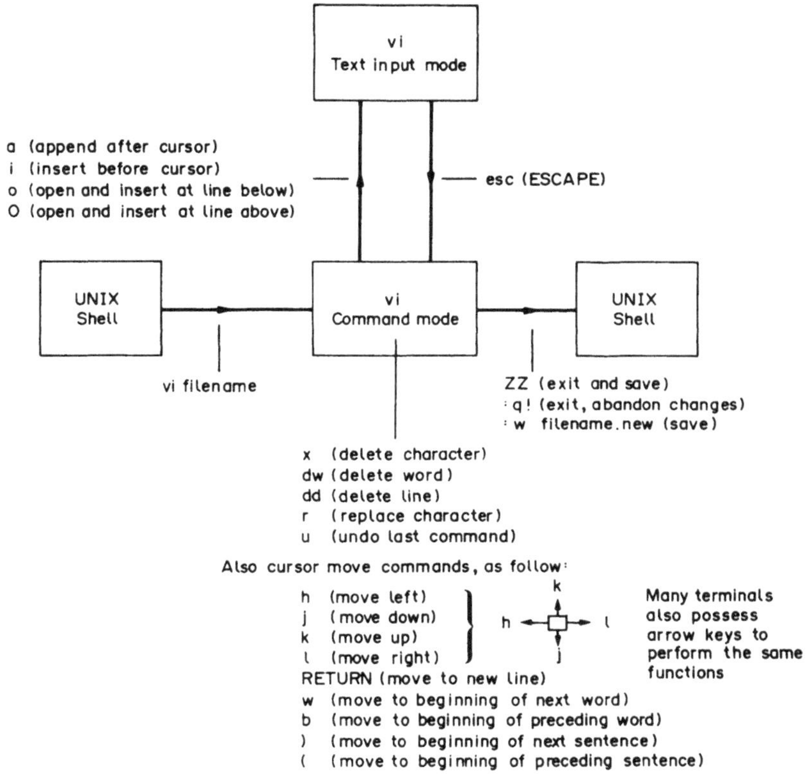

FIG. 13.2. Screen editor (vi) modes of operation.

the main commands available. The basic stages which must be obeyed in order to create a text file, e.g. a high-level language program, are:

(a) Enter vi using:
vi filename ← this is a UNIX command, of course

The screen editor vi possesses a buffer in memory into which a new file is placed, or file changes are made. When the user has finished entering or correcting his file, he must notify vi if he wishes to save the buffer contents on disc.

The screen editor responds to the command above by displaying the text file (or its first page or "windows") if the file exists, and displays the following at the bottom of the terminal screen:

"filename" 7/219 (for example)

to denote that the current file possesses 7 lines and 219 characters. The cursor is positioned over the character in the top left-hand corner of the screen.

(b) Move cursor using:
h (left), j (down), k (up), l (right) or RETURN (new line)
or the block move commands:
w (next word), b (preceding word),) (next sentence) or ((preceding sentence).

(c) Delete (if required) using:
x (one character), dw (one word), dd (one line) or r (replace).
For example, the cursor is moved to the required position on the screen using the h j k l commands above, and if:
rQ
is entered, the character under the cursor is replaced by Q.
Notice that the u (undo) command cancels the previous command.

(d) Text input mode using:
a (append after cursor)
i (insert before cursor)
o (open, moving down one line)
O (open, moving up one line)
When the user has finished inputting text, he presses:
ESCAPE
to leave text input mode and return to vi normal command mode.

(e) Quit vi using:
ZZ (writes buffer to disc, and returns to UNIX shell)
:q! (abandons buffer, and returns to UNIX shell)
:w (writes buffer to disc, remains in vi)—an argument of a new filename can be added to obtain both original and updated files if required.

These commands are the most commonly used. The screen editor possesses additional commands to justify its occasional description as a "word processor". Notice that if vi cannot obey a user command, it may flash the screen or sound the terminal bell. The following sequence of operations allows the user to abort the edit procedure:
BREAK
:q! ←——— :e! if a re-edit is required
RETURN

13.7 Sample Procedure for Pascal Program

Step 1—Log in

```
login:            ← UNIX displays
student3          ← User enters login name
password:         ← UNIX displays
gandalf           ← User enters password (not echoed)
%                 ← UNIX displays (prompt)
```

Step 2—Check if any mail (and reply to other user)

```
You have mail           ← UNIX displays (if mail exists)
%mail                   ← User enters (% is UNIX prompt)
(UNIX tells user details of mail)
p                       ← User enters (to read message)
d                       ← User enters (to delete message)
q                       ← User enters (to quite mail)
(User can send mail, or write message, to other user, if required)
%who                    ← User enters (to see other user identities)
(UNIX replies with details of other users who have logged in)
%write student6         ← User enters
Check the printer - it keeps jamming. What time is
tomorrow's meeting? ← User enters message
(UNIX displays this message on the terminal used by student6)
CONTROL D               ← User enters (to leave write)
```

Step 3—Create file (program) using screen editor vi

```
%vi exam.pas ← User enters (to enter vi and create exam.pas)
a            ← User enters (to use text input mode)
program exam;                                          ↑
(*produces an exam grade for a percentage              |
mark*)                                                 |
const    distinction=85;                               |
         merit=65;                                     |
         pass=40;                                      |
var      mark:integer;                                 |
begin                                                  |
      writeln('please enter mark');                    |
      readln(mark);                                    |  Pascal
      if mark>=distinction then                        |  program
             writeln('you have gained a                |
                     distinction')                     |
      else                                             |
      if mark>=merit then                              |
             writeln('you have gained a                |
                     merit')                           |
      else                                             |
      if mark>=pass then                               |
             writeln('you have gained a                |
                    pass')                             |
      else                                             |
             writeln('you have failed');               |
end.                                                   ↓
ESCAPE       ← User enters (to return to vi command mode)
ZZ           ← User enters (to leave vi and return to UNIX)
```

Step 4—Compile program

```
%pascal exam.pas       ← User enters (to compile exam.pas)
```

If the program compiles without any errors, the resulting executable program is named "a.out". This can be run by entering:

```
%a.out                 ← User enters
```

Each compiler (Pascal, FORTRAN and C) produces a file with the same name (a.out). Therefore the user normally renames this file, as follows:

```
%mv a.out examgrade ← User enters
```

If the Pascal compiler reports any errors, the user must correct his source program using vi, before submitting it to the compiler again.

Step 5—Examine files created and tidy up system

```
%ls -l            ← User enters (to examine his directory)
%cat exam.pas     ← User enters (to observe Pascal program file)
%pr exam.pas      ← User enters (to print file)
%rm exam.pas      ← User enters (to remove file)
%logout           ← User enters (to terminate use of UNIX)
```

13.8 Exercises

Clearly the best way to learn how to use UNIX is to experiment with it practically. Therefore the reader is encouraged to gain access to a UNIX system, and to apply as many of the commands detailed above as possible. In particular the entry procedure in section 13.7 should be attempted. The program could be altered in some way, and then re-edited and re-compiled. An alternative Pascal program that can be entered, compiled and tested is as follows:

```
program roomvol;
(*program to calculate volume of room*)
(*demonstrates use of a function*)
var
    length,width,height:real;
function volume (length,width,height:real):real;
    begin
        volume:=length*width*height
    end;
begin (*main program*)
    write('please enter room length');
    readln(length);
    write('please enter room width');
```

TABLE 13.1 *Commonly used UNIX Commands*

Command	Description
Directory control	
mkdir argument	make directory
cd argument	change directory
rmdir argument	remove directory
pwd	print working directory (gives full pathname)
ls argument	lists contents of directory (current directory if no argument is used); ls –1 produces "long" listing
File control	
cat argument	concatenate and print file
rm argument	remove file
cp argument1 argument2	copy file (from argument1 to argument2)
mv argument1 argument2	move file (rename file)
pr argument	print file
vi argument	text editor (to create a file)—screen oriented
ed argument	text editor (to create a file)—line oriented
spell argument	checks correct spelling throughout a file
Electronic mail	
mail	receiving (or sending) mail
who	who is on the system
write argument	write message to another named user
biff y(or n)	permit (block) mail
mesg y(or n)	permit (block) messages
General	
pascal argument	compile Pascal program
f77 argument	compile FORTRAN program
cc argument	compile C program
logout	leave UNIX (terminal is then ready for next user)
date	gives time and date
learn	instructions about UNIX
man	on-line UNIX manual

```
    readln(width);
    write('please enter room height');
    readln(height);
    write('volume of room is   ');
    write(volume(length,width,height):8:1);
end.
```

Table 13.1 summarises the UNIX commands described in this chapter.

Bibliography

1. *UNIX Primer Plus.* Mitchell Waite, Donald Martin and Stephen Prata. Sams, 1984.
2. *Using the UNIX System.* Richard Gauthier. Reston, 1981.
3. *A User Guide to the UNIX System.* Rebecca Thomas and Jean Yates. Osborne/McGraw-Hill, 1982.

GLOSSARY

Accumulator. A special *CPU register* that receives the results of most *ALU* operations.

A/D converter. *Analogue* to *digital* converter.

Address bus. The *microcomputer bus* that carries the *memory* address of the *instruction* that is being fetched, or a *data* item that is being transferred between the *CPU* and *memory* or *input/output*.

Address decoder. A circuit that generates *chip select* signals for each *memory* or *input/output* chip within a *microcomputer*.

Addressing mode. A method of specifying the location of a *data* item that is being accessed within an *instruction*.

ALU. Arithmetic and Logic Unit. The module within the *CPU* that performs arithmetic, e.g. add and subtract, and logic, e.g. *AND* and *OR*, operations.

Analogue. A continuous signal that can take any value over its range.

AND. The *Boolean* logic function that generates logic 1 only if both comparison (or input) bits are also at logic 1.

Application program. A *program* in a *disc*-based *computer* that tailors the machine to a specific commercial/scientific function, e.g. *word processor*, spreadsheet, database, payroll.

ASCII. American Standard Code for Information Interchange. The code that is used to represent characters in *computers*, printers and *VDUs*.

Assembler. A *program* that converts an *assembly language* program into *machine code*.

Assembly language. A programming language that is line-for-line convertible to *machine code*, but uses *mnemonics* for the *instruction* type ("opcode") and labels for *memory* addresses.

Audio cassette recorder. A domestic tape recorder that is used to store *microcomputer programs*.

Backing store. A bulk storage device, e.g. *floppy disc* or *hard disc*, for programs and data files.

BASIC. Beginners All-purpose Symbolic Instruction Code. The most popular *high-level language* that is used with *microcomputers*.

Batch file. A *file* of *operating system* commands.

Baud rate. The speed of transmisson of *serial data* expressed in bits/second.

BCD. Binary coded decimal. A 4-bit code that represents the numbers 0 to 9.

Binary. A number system that uses the base of 2. The only symbols used in binary numbers are 0 and 1.

Bistable. Two-state. A "bistable multivibrator", or "flip-flop" holds either logic 0 or logic 1.

Bit. Binary digit. A bit has two states—0 and 1.

Boolean logic. A collection of logic functions named after George Boole. The Boolean logic functions AND, *OR* and *EXCLUSIVE OR* are applied by *software* on *binary* numbers in *computers*.

Bootstrap. A *program* that loads the main program (normally the "*operating system*") from *backing store* into *memory* when the *computer* is switched on.

Bounce. Unwanted repeated operation of a mechanical contact.

Branch. As for *jump*.

Breakpoint. A stop that is inserted into a *program* to assist in the testing of a new or faulty program.

Buffer. A temporary storage *register*.

Bug. A *software* error.

Burst memory transfer. A continuous *DMA* transfer.

Bus. A set of signal connections that have a common function. A *microcomputer* possesses an *address bus, data bus* and *control bus*.

Byte. Eight *bits*.

Cache. A fast *memory* circuit that is placed between the *CPU* (or on the *CPU*) and its memory circuit. It stores the most frequently addressed locations.

Call. An *instruction* that transfers *program* control to a *subroutine*.

Carry flag. A bit in a *CPU* status register which indicates that the result of an *ALU* operation has exceeded the number range of the *ALU*.

Central processor unit. See *CPU*.

Chip. A common name for an *integrated circuit*.

Chip carrier. An IC package that uses the four sides of the device for interconnecting pins—used for devices with large numbers of interconnecting pins.

Chip select. A control signal that activates a *memory* or *input/output* chip.

CMOS. Complementary metal oxide semiconductor. A family of *integrated circuits* that offers extremely high packing density and low power consumption.

Common bus. A set of interfacing connections that allows *microcomputer* boards to be interconnected.

Compiler. A *program* that converts a *high-level language* program into *machine code* before program run time and stores both versions on *backing store*.

Complement. Change a *bit* from 1 to 0 or 0 to 1.

Computer. A programmable *data* processing system.

Concurrency. The execution of more than one function by a *computer* at the same time, e.g. parallel processing with more than one *CPU.*

Control register. The *register* in a programmable input/output device, e.g. *UART, PIO, CTC* or *FDC*, that is used to select the programmable options within the device.

Control unit. The module within the *CPU* that examines and implements the current *instruction.*

Counter/timer. See *CTC.*

CPU. Central processor unit. The main *computer* module, which fetches and implements *program instructions*. Its main sub-modules are the *ALU* and *control unit.* In a *microcomputer* the CPU normally forms a single *IC* and is called a *microprocessor*.

Cross-assembler. An *assembler* that generates *machine code* for a *CPU* that is of a different type to that used in the assembly process.

CTC. Counter-timer circuit. A programmable input/output circuit that can be used to generate timer *interrupt* pulses, generate time delays or count external pulses. A CTC is either contained within a *PIO* chip, or it constitutes a separate chip.

Cursor. A small area of light on a CRT screen (part of a *VDU*) at which characters which are entered by the operator will appear.

D/A converter. *Digital* to *analogue* converter.

Daisy-chain. A connection system in which signal connections are linked from one module to another.

Data. A general term that can describe numbers, characters or groups of *bits* suitable for processing by *computer program*.

Data bus. The *microcomputer bus* that carries *data* (and *program instructions*) between *CPU* and *memory* or *input/output.*

Debugger. A test *program* that is used to locate and eliminate errors (or bugs) in a program under development.

Denary. The decimal number system that uses a base of 10.

Digit. Each symbol in a number system, e.g. a *binary* digit can be 0 or 1.

Digital. Possessing discrete states. *Computers* operate using *binary* signals, i.e. possessing only two states.

DIL. Dual-in-line. The standard *IC* package, in which interconnecting pins are placed along two sides only of the device.

Directory. A list of all *file*names in a *computer* with a *backing store.*

DMA. Direct memory access. *Data* transfer between *memory* and *input/output* without passing through the *CPU*.

Dot matrix. A method of constructing characters using an array of dots, e.g. CRT/*VDU* and printer.

Dynamic RAM. *RAM* memory that requires a regular refresh operation to prevent corruption of stored *bit* pattern.

Editor. A *program* that allows the operator to enter or amend a text *file*.

EPROM. Erasable programmable read-only memory. *ROM* that can be erased by exposure to ultraviolet light and then re-reprogrammed.

Exception. An alternative name for *interrupt* used by some *microprocessor* manufacturers.

EXCLUSIVE OR. The *Boolean* logic function that generates logic 1 only if both comparison *bits* are different.

Execute. To run a *program*. Alternatively the second part of the *fetch/execute cycle* which is implemented when the *CPU* obeys an *instruction*.

FDC. *Floppy disc* controller. A programmable *input/output chip* that controls a floppy disc drive.

FET. Field effect transistor. The principal component in *MOS* and *CMOS* circuits.

Fetch. The first part in the *Fetch/execute cycle* which is implemented when the *CPU* obeys an *instruction*.

Fetch/execute cycle. The basic cycle that is implemented by the *CPU* when it obeys an *instruction*. Firstly the instruction is fetched from *memory* and secondly it is examined by the *control unit* and executed.

File. A *program* or *data* module held on *backing store*.

Firmware. *Program* or *data* resident in *ROM*.

Flag. A *bit* that indicates a specific condition or event.

Floating point. A number representation system for large and fractional numbers—the number is split into mantissa and exponent.

Floppy disc. A *backing store* medium that employs flexible magnetic discs.

Flow chart. The diagrammatic representation of the operation of a *program*.

Format. To initialise a blank *floppy disc* (or *hard disc*).

Gate. A *digital* circuit with more than one input, but only one output. Gates perform *Boolean* logic functions.

GPIB. General-purpose interface bus. An alternative name for the IEEE488 *common bus*.

Hard disc. A *backing store* medium that employs a non-removeable hard disc. A hard disc is faster, more expensive and possesses larger storage capacity than a *floppy disc*. It is often called a "Winchester" disc.

Hardware. The physical equipment in a *computer* (to be distinguished from *software*).

Hexadecimal. A number system that uses a base of 16. Its particular use is to represent long *binary* numbers in an abbreviated form.

High-level language. A programming language that is similar to spoken language. A high-level language *program* must be converted to *machine code* before it is executed in a *computer*.

Hit rate. The success rate of locating the contents of a *memory* location in a *cache*.

IC. *Integrated circuit.*

In-circuit emulator. A combined *hardware* and *software* system that is used in a *MDS* to test a new *microprocessor-based product.*

Initialise. To set up an *input/output chip,* e.g. a *PIO, UART, CTC* or *FDC,* to one of its programmable states.

Input/output. The *hardware* within a *computer* that connects the computer to external *peripherals* and *devices.*

Input port. A circuit that passes external *digital* signals (normally 8) into a *CPU.*

Instruction. A single operation performed by a *computer.* A *low level language program* consists of a list of instructions.

Integrated circuit. A circuit package that contains several components built into the same semiconductor wafer. This silicon wafer is housed inside either a *DIL* or *chip carrier.*

Interactive. The characteristic of a *program* that asks the operator questions during the implementation of the program.

Interface. To interconnect a *computer* to external devices and circuits.

Interpreter. A *program* that converts a *high-level language* program into *machine code* at run-time, rather than prior to run-time (see *compiler*).

Interrupt. An external signal (part of the *CPU*'s control bus) that suspends a *program* operating within a *computer* and causes entry into a special *interrupt* program. The latter is normally named an *interrupt service routine.*

Interrupt service routine. A *program* that is entered following an *interrupt.*

Jump. An *instruction* that sends *program* control to a specified *memory location.*

K. A symbol that represents decimal 1024.

Kansas standard. A signal specification for *data* storage on *audio cassette recorders.*

Label. A name given to a *memory* location in an *assembly language program.*

Language. A prescribed set of characters and symbols which is used to convey a *program* to a *computer.* A programming language can be a *high-level language* or a *low-level language.*

Latch. A circuit that staticises *bits.*

LED. Light-emitting diode. A diode that emits light when current passes through it. It is often used in a LED *segment display* unit or for single-*bit* indication.

Loader. A *program* that converts a "printable" version of a *machine code program* into executable machine code.

Logic. The application of a range of circuit building blocks to perform switching and control functions.

Logic level. The voltage value that is used to indicate logic 0 or 1. For *TTL* and TTL-compatible circuits (including most *microprocessors* and their support *chips*) logic 0 = 0 V, and logic 1 = +5 V.

Loop. A section of *program* that is executed more than once.

Low-level language. A *computer* programming *language* that specifies each operation/instruction that the *CPU* is to perform. There are two classifications of low-level language: *assembly language* and *machine code*.

M. A symbol that represents approximately a million (1,048,576). See also *K*.

Machine code. A *program* expressed in *binary* form, i.e. in the way in which it is executed within the *CPU*.

Mail. Messages that can be passed between users on a multi-user *computer* system.

Mainframe computer. A large multi-user *computer* with high processing power and a wide range of *peripherals*. Typical applications are for payroll, large customer accounts systems, large database applications.

Main memory. Fast *memory* which holds the *program* currently being executed. Main memory can be *ROM, RAM* or a mixture of the two.

Matrix printer. A printer that constructs characters using a dot matrix.

MDS. Microprocessor development system. *A computer* that is used to develop *software* for prototype *microcomputer* applications.

Memory. Any circuit or *peripheral* that staticises *data*. Normally the term is used in place of *main memory*.

Memory management. A circuit (sometimes part of the *CPU* in the case of 32-bit *microprocessors*) that can modify the address generated by the CPU.

Memory map. A diagrammatic representation of the organisation of the *memory* range of a *computer*.

Memory mapped input/output. *Input/output* devices that are treated by *hardware* and *software* as *memory* devices.

Microcomputer. A complete *computer* on a handful of *integrated circuits* (or even a single integrated circuit). *VLSI* components are used for *CPU*, *memory* and *input/output*.

Microprocessor. A *CPU* constructed on a single *VLSI integrated circuit*.

Minicomputer. An arbitrary name given to a multi-user *computer* that performs down-market applications to a *mainframe computer*. Minicomputers are used extensively for industrial control applications. They are normally 16-*bit* devices.

Mnemonic. A group of letters (or symbols) that is used to represent the function of an *instruction* expressed in *assembly language* form.

Modem. An item of equipment that converts logic levels to frequencies, and vice-versa. It is used for *serial* communication systems that pass through the public telephone network.

Monitor. The main *program* in some *microcomputers.*

MOS. Metal oxide semiconductor. A family of *integrated circuits* that offers high packing density (VLSI). Most *microprocessors* and their supporting *memory* and *input/output chips* are constructed using MOS technology.

Multiplexing. The technique of passing more than one signal along a single conductor.

Nesting. A *program loop* within another loop. Alternatively a *subroutine* (or *interrupt service routine*) within another subroutine (or interrupt service routine).

Nybble. 4 *bits.*

Object code. The name given to a *machine code* version of a *program*. The term is used to distinguish this version from the "*source program*" (*assembly language* or *high level language*).

Opcode. The part of a *machine code instruction* that specifies the function of the instruction, e.g. add, shift, *jump*.

Operand. The part of a *machine code instruction* that specifies the *data* value or its *memory* address.

Operating system. The main *program* in a *disc*-based *computer.*

OR. The *Boolean* logic function that generates logic 1 if either of the comparison (or input) *bits* is set to logic 1.

Output port. A circuit that passes *digital* signals (normally a group of 8) outside a *microcomputer.*

Parity. The number, expressed as odd or even, of 1 s in a *data* value.

Pascal. A *high-level language.*

Pathname. The full name of a *file* in a multi-*directory computer* system, e.g. *UNIX.*

PCB. Printed circuit board. A conventional circuit board with etched copper track interconnections between components.

Peripheral. An item of equipment that is external to a *computer*, e.g. printer, *VDU*, *floppy disc.*

PIO. Parallel input/output. A programmable multi-*port input/output chip.*

Pipe. An arrangement in *UNIX* whereby the input of one command is the output of another.

Pipeline. A small storage area within some *CPUs* that hold the next few *instructions* to be obeyed—the instructions are read out of *memory* before they are required in order to increase *program* execution speed.

Pixel. A dot position on a *CRT* screen.

Poll. To regularly check the status of an external signal or device by *software*.

Port. An input or output parallel-connection channel between a *computer* and external equipment. Normally a port is 8-*bits* wide.

Procedure. A section of *program* that performs a specific task—applied in *Pascal*.

Program. A set of processing steps that a *computer* is required to perform.

Program counter. A *CPU register* that holds the address in *memory* of the next *instruction* to be obeyed.

PROM. Programmable read-only memory. *ROM* that is programmed after the *chip* is manufactured. Once programmed it cannot be altered.

Pseudo-instruction. An *instruction* in an *assembly language program* that acts as a command to the *assembler* and is not converted to *machine code*.

RAM. Random access memory. *RAM* is semiconductor read/write memory. It is misnamed because *ROM* is also random access. There are two classifications of RAM: *static RAM* and *dynamic RAM*.

Read. To transfer *data* from *memory* to the *CPU*.

Refresh. To re-instate *data* in *dynamic RAM* or displayed on a *segment display* or CRT.

Register. A storage device for several *bits*. A *microprocessor* contains several work registers, as well as special-function registers.

Return. An *instruction* that returns *program* control to a main program from a *subroutine* or *interrupt service routine*.

RISC. Reduced instruction set compuer. A *CPU* that possesses only a limited *instruction* set in order to achieve high-speed performance. An example is the *transputer*.

ROM. Read-only memory. *ROM* is semiconductor *memory* which can only be read. There are three classifications of ROM: ROM, *PROM* and *EPROM*.

RS232-C. The internationally recognised specification for *serial data* transfer between *computers* and serial-drive *peripherals* (or other computers).

SBC. Single board computer. A complete *microcomputer* circuit on a single *PCB*.

S-100 bus. A *common bus* applied in multi-board *microcomputer* systems.

Segment. An area of *memory*—typically 64K in several 16-*bit microprocessor* systems.

Segment display. A display that constructs numbers and letters by a network of segments.

Semiconductor memory. *ROM* and *RAM*.

Serial. The transfer of *data* items by setting one *bit* at a time on a single conductor.

Shift. Transfer of *data* to the left or right (normally within a *register*).

Software. *Computer programs* and *data files.*

Source program. The name given to an *assembly language* or *high-level language* version of a *program.*

Stack. A reserved area of *memory* (*RAM*) that is used by most *microprocessors* to store the return address in *subroutines* and *interrupt service routines.*

Static RAM. Conventional *RAM*, unlike *dynamic RAM* which requires refreshing.

Status register. A collection of *flag* bits in a *microprocessor* that indicates the state of the *ALU.*

Structured programming. The technique of designing a *program* in modular form, with the intention of making the program easier to understand, test and modify.

Subroutine. A section of *program* that is separated from the main program, but can be called several times from the main program.

System program. The *operating system* or a utility *program* that supports the operating system, e.g. *editor, compiler, assembler.*

Tag address. The address of a *memory* location that is held in a *cache.*

Transputer. A *microprocessor* family produced by Inmos. A transputer device is designed to operate in a parallel array of transputers in order to perform the *CPU* function (using the principle of *concurrency*).

Tristate. A circuit in which its outputs can be set into one of three states: logic 0, logic 1 or "floating" (high-impedance state, i.e. electrically isolated).

TTL. Transistor transistor logic. A family of *integrated circuits* that preceded *MOS* and *CMOS*, but is still widely used for *gates*, buffers, flip-flops (*bistables*), etc.

TTY. Teletype. A name sometimes given to a printer.

Two's complement. A *binary* numbering system used to represent both positive and negative numbers—the most significant *bit* acts as a sign bit.

UART. Universal asynchronous receiver transmitter. An *input/output chip* that handles *serial data* transfer, e.g. to *VDU*, printer or other *computer.*

VDU. Visual display unit. An operator device that includes a CRT for display purposes and a keyboard for manual entry.

Vector. Part (or all) of a *memory* address that contains the start address of an *interrupt service routine.*

Virtual memory. Main *memory* that is extended to include *backing store* as if it is an extension of main memory.

VLSI. Very large scale integration. A measure of the packing density of an *integrated circuit* (more than 1000 *gates* per *chip*). Used as a description of *MOS* and *CMOS* devices.

Volatile memory. *Memory* that loses its stored *bit* pattern when power is removed.

VRAM. Video RAM, i.e. *RAM* that is used to store video *data.*

Winchester. Another name for a *hard disc.*

Word. A unit of *data* in a *computer*. The word length is the same as the bit length of the *CPU*, e.g. *microprocessors* are expressed as 8-*bit*, 16-bit or 32-bit devices.

Word processor. A *program* that is used to create and amend text *files*, e.g. letters, reports.

Write. To transfer *data* to *memory* from the *CPU.*

Write protect. To set a *backing store* device to read-only to protect against over-writing.

INDEX

Printed in Poland
by Amazon Fulfillment
Poland Sp. z o.o., Wrocław

74934428R00108